# True Time Travel Stories

## Amazing
## Real Life Stories
## In The News

### Compendium Containing:
### Book One & Book Two
### Of The Time Travel Series

Richard Bullivant

# Book One

# True Time Travel Stories

## Amazing
## Real Life Stories
## In The News

# Contents

# Introduction

**Has the human race always been fascinated with the possibility of time travel?**

We know for certain that humans have been dwelling on the subject of time itself for at least 70,000 years and possibly longer.

We can say this because cave paintings rendered by the aboriginal people of Australia dating to at least 70,000 BC show themes connected to time. Cave paintings in Europe dating back more than 30,000 years also bear symbols indicating perceptions associated with time. It's therefore clear that our ancient ancestors were thinking about time.

Philosophers such as Ken Wilber say that prehistoric people began to 'obsess about time' when human consciousness developed to the point where it became 'self-reflective'. In other words, at some point, people began to think: *'Hey, I'm somebody! I'm a unique individual. I'm not just part of a herd or an extension of nature. I'm here for a certain amount of time.'*

The trouble with finally achieving self-reflective knowledge, however, is that people also realized that their 'time' on earth was limited; that is, they knew they were going to die at some 'time'. And since no one wanted to die, they began to worry about the passage of time. They understood that when their 'time ran out' they would be dead.

## Time as an Actual Substance

It didn't take long for time itself to become a kind of commodity - an actual *thing* - and you always wanted more of that *thing* that was called time.

Furthermore, after time became *something* it was inevitable that people began to wonder if they could manipulate it in some way - perhaps even travel in it or upon it, perhaps like a flowing river. And maybe you could paddle both *upstream* and *downstream*.

But what is *downstream* in terms of time? Well, this brings up a second factor in the development of human consciousness: The advent of memory. Now people could 'see in their minds' that place where they existed yesterday. They could remember what they were doing a week ago, or years ago. For example, they remembered how they survived the blizzards and food shortage of that past, long ago bitter winter. This created the idea that a certain amount of time extended backwards - *downstream*.

Now they knew that there was a mysterious *future time* out there - *upstream* - and that in some location on that future upstream time, they would die. They could combine this with a *past time* as perceived by memory.

## Centuries of Debate

It is interesting to note that human beings have been debating the concept of time since almost the beginning. Many ancient Greek philosophers argue that time does not exist, but is merely an illusion of the mind. It's true that, in general, the Greeks thought of time as some kind of vast system of cycles, but brilliant men such as Plato argued that time was merely a metaphor and not something that should be con-considered as real.

In the 4^(th) Century AD, the highly influential philosopher St Augustine put forward his idea that time was *an illusion of the mind.*

Other cultures and philosophies, most notably Buddhist thought, contend that the 'eternal present' is the only thing that exists. They say that past and future are illusions. They point out that even when you use memory to think about past events, you are not actually travelling into the past to get those memories - you are accessing them in the present. Your memory only gives you the illusion of a past time.

Now fast forward to modern-day physics. Even the likes of the mighty Albert Einstein said in a letter to a fellow physicist: *'...for us physicists believe the separation between past, present and future is only an illusion, although a convincing one.'*

So it seems that Einstein believed the Buddhists got it right on this one. Today, most physicists probably agree that time is more illusion than real - and yet it's not that simple. That's because many scientists contend that time is real, but only when considered as a component of space. They say we can never talk about time without also talking about space, and so we have the now popular scientific term *space-time.*

**What the Scientists Say ....**

It's fair to say that most scientists today will tell you that time travel is impossible. Three of today's top physicists - Charles Liu, Brian Green and Michio Kaku - all hold that time travel is, if not impossible, unlikely in the extreme. However, one of the most brilliant minds of our time, physicist Stephen Hawking, disagrees - although only partially. He believes that time travel is theoretically possible, but only into the future.

**.... But What Do Real People Experience?**

The opinion of science, however, has never stopped thousands of people around the world from reporting what they firmly believe are actual experiences of spontaneous time travel! Still others insist that time travel is not only possible, but they have already done it as part of top secret government programmes.

Claims for time travel range from the highly flaky to the astoundingly believable. They are especially difficult to dismiss when time travel reports come from absolutely ordinary, rock-solid people who have nothing to gain by proclaiming they travelled in time.

Many people who report time travel experiences don't necessarily believe it themselves. What happened to them was so strange, so unexpected, yet so real; they simply have no other good explanation for their experience.

You will meet a number of such individuals in this book, most of their stories straight out of the headlines of local newspapers. No doubt a story or two will strike the reader as pure balderdash. On the other hand, some of these cases of time travel are tantalizing and unexplainable. They also come with a certain amount of solid evidence, such as stopped clocks, frozen machines and electromagnetic devices acting in inexplicable ways.

Physicist and NASA scientist Tom Campbell said that scientific advances always 'come from the fringe.' Thus, even if you consider some of these stories stepping dangerously 'out there' onto that fringy edge, remember that many of yesterday's fringe theories are today's scientific fact. At the very least, it doesn't hurt to approach the idea of time travel with an open mind and a sense of wonder.

# The Man Who
# Froze Time

One of the strangest cases involving the manipulation of time was first reported to the world in a 1977 Vancouver Sun Times article. The story was picked up by other media around the world and created something of a sensation. The events surrounding this intriguing time manipulation tale have also been featured or referenced in dozens of books since.

A large body of circumstantial evidence seems to suggest that a humble appliance repair man by the name of Sid Hurwich of Toronto, Canada, had invented a device that could actually freeze - or change - the flow of time in any particular location where the mechanism was placed. But the device would also appear to have the ability to *send out beams of influence* to manipulate objects in distant locations.

If all this sounds too sensational to believe, consider the fact that Hurwich, a Jewish man living in Canada, was awarded one of *the Protectors of the State of Israel Awards* after providing his device and its schematics to the Israeli Department of Defense.

Many believe the Israelis used the Hurwich time-altering device to pull off a number of daring military operations, including the astounding raid on Entebbe on 3 July 1976, during which Israeli aircraft somehow managed to avoid the radar detection of six nations - a feat that should have been impossible.

**Sid Hurwich – A Man of Unusual Ability**

Sid Hurwich was an ordinary appliance repair man with no college education, but he possessed a knack for tinkering and figuring things out. He was born in 1918 in Toronto and by age nine, his daughter said that Sid was already collecting whatever random junk he could find and would then assemble it into useful appliances. Hurwich loved to take things apart to see how they worked. He was obsessed with finding out how things ticked.

In 1934 Sid earned the distinction of being the first private appliance repairman in all of Canada. Prior to that time, only company trained technicians could repair household appliances. But if anyone needed something fixing, they knew that Sid Hurwich was the man to call. By World War II he was known far and wide as *the man who could fix anything*. So valuable was Sid that his local power company, Ontario Hydro, pulled strings to keep him out of the army so that he could help develop the infrastructure of the public electrical utility.

After the war Sid set up his own company, Shock Electric, which became a success. He established a second business, SidCo Company, which built electronic parts and which became an even bigger success. Although still a young man, he suffered a heart attack in 1950, sold his companies for an enormous sum of money, perhaps millions, and retired.  He was thirty-six years old.

## Catch a Thief in a Time Trap?

In 1969 Sid Hurwich's home city of Toronto was plagued by a rash of bank robberies.  The police and public were on edge.  Fortunately, as it happened, Sid knew some of the local police.  After reading about the spate of bank robberies Sid said that a brilliant, but simple idea popped into his mind.  So he called up the police and told them something astonishing: He had figured out a way to capture bank robbers using one of his latest inventions.

Since this was the highly respected, financially successful businessman Sid Hurwich, the police were willing to listen.  A police inspector by the name of Bill Bolton visited Sid at his home, together with two fellow police officers.

The article in the Vancouver Sun Times gives this account:

*'All I recall,' said Bolton, 'is that it was under the table - the device, whatever it was - and there was a bedspread over the table.  He froze my service revolver.  You couldn't pull the trigger, you couldn't lift it up off the table and even on the table you couldn't pull the trigger.'*

*Sid said: '....Now take a look at your watches.'  I remember one of them said, 'When did this happen?' and I said, 'The minute you walked through that door. You walked in there about 25 minutes ago.  Now look at your watches. You're late by about 25 minutes.'*

*As the security officers filed out of his home, Sid's wife overheard one of them suggest that the army should be told about the device.*

*'That was the first time it entered my mind for war or army purposes or anything like that,' Sid said.*

He went back to work in the basement. When he felt the device was ready, he contacted his brother living in Israel. Sid received a visit shortly thereafter from two high-ranking Israeli officers. After a brief demonstration, they walked out with the working model, and every plan and design Sid had.

But is there any real evidence that the Israeli military put the Hurwich time altering device to constructive use? There is. The evidence comes by the way of a British publication called *Foreign Report* which is produced by the highly prestigious Economist magazine. In a story about the famous Israeli raid on Entebbe, the Foreign Reports says:

*'...all that could be learned officially was that Hurwich's invention was used in the Israeli raid on Entebbe last year.'* (The source was quoted as being 'a high-ranking Israeli military official'.)

The article goes on to say:

*'The device sends out electronic rays to alter the natural composition of electronic fields and centres of gravity of weapons, instrument dials and mechanical devices ... (using) the Hurwich principle there was no reason why the new beams could not reach and disable tanks, ground-to-ground missiles and complete radar systems. The beams could also be tacked to-*

*gether to form a screen that would make whole zones safe from bombs and missiles.'*

## Secret Time Weapon Still in Use?

It's interesting to note that since the Entebbe raid in 1977, the Israelis have pulled off similar feats a number of times, including the bombing of what they believed to be a nuclear weapons development plant in Iraq in the 1980s during the reign of Saddam Hussein. Iraq had excellent radar detection and anti-aircraft weaponry, yet the bombing was over and done with before the stunned Iraqis realized their airspace had been breached.

And just recently, in January of 2013, Israeli fighter jets entered Syrian airspace to bomb a convoy of anti-aircraft missiles near the Syria-Lebanon border. Immediately after the Syrian mission, many pundits expressed surprise at the boldness of the Israelis, citing the fact that Syria has an effective radar system and anti-aircraft missile batteries that could have posed a grave threat to the Israeli pilots. Many pundits have asked: 'How did these fighter jets go undetected into Syrian airspace?'

Does the Israeli military still have the Hurwich device, and are they still using it on 'special occasions' or in dire situations of military crisis?

Sceptics say that if they really had a device that could provide a kind of shield from bombs and missiles, freeze weapons, and even slow or halt time, the Israeli military would be invincible and would not have suffered as many casualties as it has in its decades long struggles with its antagonists in the Mideast. Others counter, however, by saying that the Hurwich device may be a two-edged sword. If it can 'freeze' the weapons of the enemy, it would also freeze the weapons of those employing the device. Thus, the device can perhaps only be used in a limited way - enough to knock out radar detection and other enemy electronics, for example - but not to induce the full-blown time-freeze affect for the benefit of the user.

**Whatever Happened to Sid Hurwich?**
While many people have written and speculated about the time altering device of Sid Hurwich over the years, there has never been a follow-up story since he appeared in the Vancouver paper back in 1977 - although the device and the implications surrounding it have been discussed in a number of books, especially two popular books written about the raid on Entebbe.

But many wonder: What would Sid Hurwich say about his device if he were alive today? And for that matter, whatever happened to Sid Hurwich?

It is likely that by this time Mr Hurwich has passed on. Born in 1918, he would now be near 100 years old. He was aged 64 in 1977 when his picture appeared in the Toronto newspaper. However, locating even so much as an obituary or any information about him has frustrated a number of investigators - many of whom are interested in speaking with him in the hope that he might share his secrets about how his time influencing invention worked.

Many are still tantalized by what he told police investigators about his device: 'It's not really a new invention. It's designed on principles that are already well known. I just thought of it one day. When I heard about the bank robberies, I knew this could work.'

But Sid Hurwich has long since receded into the background after that initial burst of fame in the late 1970s. His greatest invention may still be in the hands of the secretive Israeli Department of Defense. If the Hurwich device truly is capable of altering the fabric of space-time itself, it remains a military secret today, and is likely to stay that way.

# Time Storms

Whirling vortexes are observed everywhere in nature. The most dramatic examples are tornadoes, hurricanes and cyclones. But we also witness whirlpools forming naturally in water, and even the stars of our vast universe form swirls - billions of stars spinning around what astronomers say are black holes at the centre of galaxies.

On a blustery day, perhaps you have seen a quantity of leaves caught up rotating in a burst of wind, or a dust devil twirl in a dusty spot. Vortexes can even form in fire - they are called plasma vortexes.

It's clear that nature loves to form vortex structures, so it may also be natural for vortexes to develop using something more interesting - the very fabric of space itself.

For example, what if a vortex formed using space, time, gravity and electromagnetic energy? This may be happening, and when it does, what you get might be a *time vortex* - or *time storm*.

There are literally hundreds of well-documented cases of ordinary people going about their daily lives when suddenly they encounter a strange mist or a swirling mass of light energy. If they become engulfed by such a phenomenon, strange things happen. These include slips in time, missing time, sudden transportation to other locations, extreme psychological disorientation - even strange visions and visitations by exotic beings.

Let's take a look at a few famous examples.

**A Soldier's Strange Story**

In 1977 the San Antonia News of Texas picked up a story that had been plastered all over newspapers in the South American country of Chile. It was about the bizarre events encountered by Corporal Armando Valdez, a soldier in the Chilean army. He was on routine patrol from where he was based at a remote military outpost called Pampa Lluscuma, near the city of Putre, Chile. The area is in the region of the high Andes.

It was 4.30am on the morning of the 25 April. Corporal Valdez was in charge of a group of six men who, nearing the end of their night patrol, were resting around a campfire and basically trying to stay awake through the rest of their shift. They were singing songs and drinking coffee when suddenly one of the men out on watch duty came running back to the campfire. He was obviously frightened. He told his comrades that he had seen a strange sight - two violet amorphous orbs or shapes of light had descended into the area. They each had red spots on their outer edges and were bright enough to illuminate the surrounding area.

Corporal Valdez ordered his men to douse the camp fires, put their weapons at ready and be on alert. In the meantime, he left the campsite to investigate. He was gone for about 15 minutes but when he returned to his men, they immediately sensed something was wrong.

Corporal Valdez seemed dazed and incoherent. But what was even stranger is that Armando had left clean and close shaven and returned with a beard! His clothes were rumpled and dirty, as if he had been out in the field for some time.

His men asked him what happened, but the Corporal only had something bizarre to say. He said this to his men:

'You will never know who we are, nor where we come from, but we will return again.'

With that, Corporal Valdez fell dead away to sleep. The rest of the men were perplexed to say the least. As their patrol leader slept, one of the men happened to notice his watch - and was surprised to see that it read 7.00am - and that the calendar on his watch now read 30 April, five days ahead of the current date! Again, Corporal Valdez had only been gone fifteeen minutes.

After two hours of sleep, Armando woke up and seemed to return to a normal state of mind, except for being mildly perplexed. He appeared confused and said to his men, 'I don't remember a thing since I walked away from the camp just now.'

The case of Corporal Armando Valdez has long been considered to be a case of UFO abduction and has been treated as such by UFO investigators. But over the years, Valdez has always denied that he was abducted, or saw any aliens. He remembers nothing at all about his missing time incident.

In the decades since this 1977 episode, the case has been revisited and studied intensely by a number of UFO investigators. One of them, British author Jenny Randles, has argued that the best explanation for the case is that Corporal Valdez encountered *a time storm*. She suggests that the violet lights encountered by the Chilean soldiers were a kind of 'atmospheric phenomenon' but in this case involving an anomaly in space-time, a naturally occurring vortex in the very fabric of space itself.

Jenny Randles submits this as the best explanation for the growth of Valdez's beard and the advanced date on his watch. Since Valdez otherwise remembers nothing, Jenny said he was simply swept up by a localized time storm - his interval within the vortex would seemed to add up to five days relative to his own experience - but only fifteen minutes to the rest of the world.

What Jenny Randles and others have no good explanation for, however, is the haunting statement of Armando Valdez: 'You will never know who we are, nor where we come from, but we will return again.'

In a recent interview Armando Valdez said he is finally preparing a book on his experience, which he hopes will tell the story from his own perspective. No release date of the book has been announced.

**Caught on Video**

Stories abound about sightings and encounters with *time storms*, but in this day and age when there seems to be a surveillance camera on every corner and everyone carries a video camera in the form of a smartphone, it seems that at least one *time storm* should have been captured on video by now.

Well, that just may be the case. In 1996 the news department of a local Florida TV station received an anonymous video taken by a security camera at a small factory in Florida. The video was actually the result of multiple cameras feeding the same deck of monitors.

The footage shows a man walking towards the rear gate of the factory. He appears to be looking for something. Suddenly a fuzzy white glow, like an electrical cloud, enters the scene and moves to intercept the man. Apparently, electromagnetic interference disrupts the video signal for a few seconds. When it clears up, the man has disappeared.

The video was subjected to painstaking frame-by-frame analysis by a team of experts, including Ted Williams, a physical scientist, John Carpenter, a psychiatrist, Dr. William Schneid, a criminologist and computer analyst Dan Ahrens. All of these men volunteer their time to investigations supported by the Florida chapter of MUFON, the Mutual UFO Network.

Based on the key frame, the team could only conclude that the man simply appeared to have vanished just as the white fog enveloped him.

The time of the incident was 11.16pm. A security guard observing the video monitors had seen it all, and went out to investigate. He was however alarmed when it became obvious that the man, a factory employee, was nowhere to be found. A search was conducted, but the man was found to be nowhere on factory grounds. Maybe he just walked off the job? They called his home, and his family reported that he was not there and should be at work at the factory.

About two hours later, at 1.06am, the fuzzy white glow once again made its appearance on the security cameras. This time, sure enough, it appeared to deposit the missing factory worker in the same spot from which he had vanished.

The video footage shows the white cloud and then, in a fraction of a second, the man reappears, but now down on his hands and knees, obviously in distress, disoriented and vomiting. A security guard runs out to help. He asks the worker what had happened, but his mind and memory were a blank. He was sent home in a state bordering on shock. He called in sick the next day, but never returned to his job again.

The MUFON team spent hours examining the video, looking at every frame, every fraction of a second, but could find no evidence of fraud. They also subjected the video to a variety of tests to rule out special effects manipulation, but none could be found.

Was he taken up by a time storm?

## The British Countryside – Time Storm Central?

Perhaps no other area of the world has more reported time slips and time distortion reports than Great Britain. Cases of vanishings and reappearances by bewildered people from *lost time* episodes date back centuries. Such events in the British Isles and

Ireland have long been associated with fairy phenomenon. It is a standard point of legend that any person foolish enough to step inside a 'fairy ring' can expect extremely weird events - one of the primary of which is missing time - and these also have often been associated with glowing clouds, mists, fogs and travelling orbs which envelop people and cause them to vanish.

**Fast forward to Modern Times.**

A case that is considered by many to be the most important UFO abduction case in British history involved an East London family who were travelling in their car near the small town of Aveley in West Essex. The event took place in October of 1974.

John Day, his wife Elaine, (not their real names) and their three small children were travelling down a lonely country lane, returning to London from a visit with Elaine's parents. Two of the children were asleep in the back seat and one was awake. At about 10pm the travellers spotted a strange oval object, which they described as a 'pale blue iridescent light' floating over the English countryside about 500 yards from their car.

The family was somewhat unnerved. The object seemed to keep pace with their car as they continued to drive east. Eventually, however, they lost sight of the glowing orb. They concluded that they had seen a UFO, and thought that would be the end of it - but the strange events of the evening were not over.

A few more miles down the road, John Day drove his car around a curve and ahead they saw a strange green fog on the road. As the car neared the fog, John and Elaine said they began to 'feel very strange.' They also noticed that they could no longer hear the engine of their car, or the sound of the tyres against the road. The car's radio erupted into loud crackling static and began to emit smoke. It seemed a small electrical fire was breaking out in the dashboard.

John reached under the dashboard and pulled the wires out from the radio - he was still driving - and a minute later they were upon the green fog and plunged straight into it. Immediately upon entering the fog the car jerked violently, and then lurched as if hitting a large bump, although they kept moving forward. The feeling inside the fog was eerie - in fact their entire world seemed to have gone silent. The view all around was hazy and shimmering as John and Elaine felt a tingling sensation in their bodies.

In what seemed like just an instant, the car emerged out of the other side of the fog. Passing through the mist seemed a mere fraction of a second. They immediately noticed something unusual, however. The car seemed to have leapt forward about a mile!

John and Elaine were extremely familiar with this road, having travelled it numerous times. They knew each tree, fence and landmark by heart. Somehow, they had skipped over at least a mile of highway. Slight feelings of disorientation continued until they finally reached their home - where they were in for yet another shock.

That's because it should have been about 10.30pm but the clock in their home read after 1.00am! Bewildered, they realized that just a second in the green mist had carved three hours out of their lives!

The Day family case incorporates many elements of both time shift and UFO phenomenon. Remember that just preceding their passage through the time storm, the Days had a long look at a large UFO that appeared to follow them. In the days that followed, the missing three hours haunted them, even though they had no conscious memories of anything strange that might have occurred in those missing hours. However, John would eventually undergo hypnotic regression, which revealed an elaborate tale of an encounter with strange beings. It was a classic, full-blown UFO-alien abduction scenario.

Both John and Elaine also experienced transformation in their personalities - mostly for the better - but not before John suffered a nervous breakdown a couple of months after his encounter. When he recovered, however, he emerged a more creative and artistic person. He left his job as a carpenter and general handyman to work with handicapped children. He had a greater sense of feeling, caring and empathy for his fellow human beings.

The larger implications of the UFO-alien abduction scenario of the Day family are beyond the scope of this book. Since their experience in 1974, the Day family have been the subject of many newspaper articles and investigations by professional and amateur ufologists. However, as the concept of *time storms* and theories about vortexes of space-time anomalies have garnered greater interest, the story of John and Elaine has increasingly been offered as the best explanation for their encounter.

**The Oz Factor**

Up until recently, events of missing time have almost always been associated with UFO and abduction phenomenon. However, researchers have many cases of missing time which seem to involve no aspect of UFO activity. There are hundreds of reports of ordinary people going about their daily lives who encounter strange clouds, mists or fogs which leave them out of place - sometimes miles distant from where they were - and displaced in time. Sometimes a few minutes or hours are missing, and sometimes it is days or months.

Entering a time storm or vortex also produces what author Jenny Randles has dubbed *The Oz Effect*. This refers to the tremendous feeling of unreality and strangeness experienced by those who are caught up in time storms. The symptoms are numerous - confusion, tingling feelings, a sense of being in a dream, a feeling that time has stopped, total silence, bizarre thoughts and visions - and the effect tends to last for days, months or years after the event. Many people also have their personalities altered, sometimes for the better, other times not.

Today, thanks to advances in physics, quantum theory and new models of just how our universe is structured in terms of dimensionality, many cases of

UFO related missing time events might better be explained by natural disruptions in the fabric of space-time. Much like storms can form in our atmosphere or in bodies of water, perhaps the very essential medium of space itself can display similar formations and, when human beings encounter them, strange things are sure to follow.

# Andrew Basiago

In 2004, a highly respected and well-known Seattle attorney started talking to the major media in his area. The story he had to tell was beyond astonishing and beyond strange. Andrew Basiago told reporters at the Seattle Post Intelligencer newspaper, and its media sister station KOMO4 TV, that when he was a child in the late 1960s he had been enlisted into a super-secret government programme involving time travel.

Andrew said his career as a traveller to other points in time and to other centuries was part of a US Government Cold War research programme dubbed *Project Pegasus*. He said he made these 'leaps in time' dozens of times, including six trips back to the days of Abraham Lincoln. He said he was present at the Gettysburg Address and that he was in the Ford's Theatre on the night President Lincoln was assassinated.

He was just six years old when his attachment to Project Pegasus began in 1967-68. His father, Raymond Basiago, was an electrical and aerospace engineer who worked on the project. Basiago said that his father had 'been ordered' to make his son a part of the project because scientists wanted to test the effects time travel would have on children as well as adults. It seems that the elder Basiago had no choice but to comply. In fact, Andrew Basiago says he was one of about 140 children in the Pegasus Programme along with about 60 adults - they were dubbed 'chrononauts'.

## Strange Beginnings

Andrew Basiago told reporters that when he was a child he had been placed into a special advanced learning programme by his father. As an ordinary first grader, Andrew simply assumed that all kids went to school this way. But Andrew's lessons were a special kind of 'rapid learning' system facilitated by special data-loading systems that were designed to allow for fast delivery and rapid absorption of massive amounts of information.

He was also asked to sign a 'loyalty oath' to DARPA, the Defense Advanced Research Projects Agency, which is part of the US Department of Defense. Project Pegasus was being conducted under the auspices of DARPA.

In late 1968 or early 1969, Andrew went with his father to a facility managed by Curtis-Wright Aeronautical facility in Wood Ridge, New Jersey, which is near where the Basiagos lived. Here he was introduced to a variety of startling and exotic technologies being developed by cutting-edge Defense Department scientists. The primary technology Andrew describes was a time travel device based on a design of electronics genius Nikola Tesla. This machine was called the *Tesla Teleporter*.

'The machine consisted of two grey elliptical booms about eight feet tall, separated by about ten feet, between which a shimmering curtain of what

Tesla called *radiant energy* was broadcast,' Andrew told the Huffington Post in April 2012. 'Radiant energy is a form of energy that Tesla discovered that is latent and pervasive in the universe and has among its properties the capacity to bend time-space.'

But there were other systems as well, such as something he calls *the chronoviser*. This was a time travel mechanism that could send a hologram-like projection of a person into another time. Andrew said a person projecting into the distant past would not have solid form while travelling via the chronoviser.

'They told us that if you get stepped on by a dinosaur or if a Viking runs his sword through your midsection, you won't be harmed because it would all just pass right through you,' Andrew said in an Internet radio interview. 'So we weren't actually there physically in another location, but a holographic analogue of ourselves were being projected into those target locations.'

He said that the real goal was for a person to travel physically with one's whole and 'normal' body into another time period and this is what the Tesla Teleporter could accomplish. He explained that stepping into the device was like 'entering a shimmering tunnel of light'. He said you would feel yourself tingling and sparking for a while and when you 'solidified again' you would be at another location in time.

**First Trips**

Although Andrew has conducted dozens of radio interviews since the initial coverage he received in the mainstream media, the specific details of how he made his first time travel journey are somewhat sketchy - or at least so voluminous they are difficult to put together. However, he says that between 1969 and 1972 he was sent on a variety of assignments, one of the primary of which was five or six trips back to gather information about Abraham Lincoln or, more specifically, information about his assassination on 14 April 1865.

For this mission, Andrew said he was attired in period clothing. His handlers at Project Pegasus, working from historical records, knew that he would be greeted at the door of Ford's Theatre by two men sitting at a table who were taking tickets. The seven year old Andrew was instructed to tell the men that his parents had gone in ahead of him and had already given them a ticket. He said they waved him through every time under this deception.

But why was Project Pegasus interested in sending a seven year old boy to witness the assassination of President Lincoln? Andrew said they wanted to determine for certain that it was John Wilkes Booth who had actually pulled the trigger, and not someone else.

He said his handlers suggested that the historical accounts of Lincoln's assassination might not be the true story, and he assumed they had reason to believe so. They seemed to think that the killing of the president might have been an inside job, possibly a government coup, and this suggested that the American government had been illegally hijacked from that moment on. They suspected that even Mrs Lincoln herself may have pulled the trigger! Whatever the case, they wanted to know.

Andrew said he tried to gain access to the balcony box seats where President Lincoln and his wife were seated, but was never able to do so - and this led to a significant problem. For example, the first time he entered Ford's Theatre and made his way to Lincoln's box seats, he was chased away by a guard. On his next trip, he found the door to the box seats locked. Each time he tried to gain close proximity to the Lincolns, he was thwarted in some way, but always in a slightly different manner.

In fact, Andrew noticed that even though he was being sent back to the same point in time on multiple assignments, something had changed every time he made a new trip. There were slight alterations in detail. Sometimes there would be a guard at the door of Lincoln's box seats, at other times not. On one occasion, he saw President Lincoln and his wife stroll through the lobby of the theatre just as he arrived, and other times they were not there.

These alterations in detail might be explained by what theorists call the Multiple Universe Scenario, Andrew contends or, more specifically in this case, multiple time lines.

Put simply, physicists have long theorized that the universe we live in is not the only or singular universe, but rather one of an infinite number of alternate or parallel universes that make up all of reality. Each universe is shifted slightly from the next. If a person somehow slips, say, into an alternate universe tucked in right next to ours, the differences between the two universes would be almost unnoticeable.

However, if a time traveller or inter-dimensional traveller veers ever further from his or her 'host universe' the alterations become more pronounced. Thus, Andrew contends that each time he was sent back to Ford's Theatre in 1865, he was landing in a slightly shifted time line, making small details different. In fact, Andrew contends, this is what caused the managers of Project Pegasus to eventually stop sending people back into the past.

Though he visited Ford's Theatre at least five times, Andrew said he was never able to witness the assassination itself, although he was close enough inside the theatre to hear the gunshot and the commotion that followed.

Andrew Basiago had one other trip into the Lincoln era, however, and that was on 19 November 1863. The target site of Lincoln's famous Gettysburg Address given in Gettysburg, Pennsylvania, several months after the Union victory is one of that war's bloodiest battles. For this event Andrew was adorned in the attire of a Union Army bugle boy - and he claims that his presence there was captured in a photograph by famous Civil War photographer Josephine Cogg. The photo shows the blurry image of a young boy standing somewhat apart from a crowd of people present at the Gettysburg Address. The picture clearly shows a boy of about Andrew's age, but is not recognizable as him, which he readily admits, but insists that it is he in the photograph.

He also says he was supplied with a letter in his trip to Gettysburg addressed to then Navy Secretary Gideon Welles to appeal for 'aid and assistance' in the event that he might be arrested - an eventuality he never encountered.

## Travels to the Future

Andrew also says he was sent into the future, to the year 2045. This was now later in his life, in the 1980s, when he was an adult. On this journey, he claims to have been sent to a facility which contained a significant store of archives of historic events dating back to the 1970s. In effect, his superiors wanted to gather all the information they could about what would be happening over the next 50 to 75 years so that the government could conduct 'contingency planning' to prepare for what was to come.

Andrew is not sure where the facility was located, or in what city, but says he is certain it was somewhere in the American southwest based on the climate he observed. He also described the world of 2045 as somewhat Utopian. The facility to which he was sent was in an urban location that appeared to be expertly integrated with the natural environment. He said the people looked pleasant and healthy, and 'seemed more generally intelligent than average people today.'

## Travels to Mars

Perhaps the strangest of all of Andrew Basiago's claims is that he also made several trips to the planet Mars when he was a young man of nineteen. These were not time travel excursions, however. Rather, Andrew says that the same kind of teleportation technology that could be used to leverage time travel can also be used to send people from Point A to Point B in the blink of an eye. It's a form of 'quantum teleportation,' Andrew says.

Again, Andrew's story is extremely detailed and complex, so what we are providing here is merely the broad framework of those events as he described them.

By the time he was nineteen, Andrew said that the managers of Project Pegasus had used a variety of techniques, from hypnosis to drugs, to erase his memory of the time travel experiences in his youth. At this point, however, he was brought back into the programme for a special assignment. By this time scientists had made considerable advances in time travel and teleportation technologies, he said.

It seems that the Mars operation was being conducted by the CIA, the American Central Intelligence Agency. Not giving him much choice in matters, Andrew was told that he was to be teleported to the planet Mars. He was incredulous, but there was little leeway for debate or argument. He immediately understood that he was not being *asked* to go to Mars as a volunteer, but that he would be going to Mars because they wanted him to - period.

His first thought was that he was to be sent to Mars in some kind of advanced conventional spaceship, and that the trip would take weeks or months of space travel. Remember, at this point in his life, he had no memory of his previous time teleportation experiences. But his CIA handlers said that all he had to do was step into an elevator-like room. When the doors closed, he would feel some moderately strange effects, and when the doors opened, he would be on the planet Mars, within minutes.

Andrew found this patently absurd but, again, he was in no position to argue or refuse anything he was being told to do. His preparation for a trip to Mars consisted of less than an hour of informational briefing, little of which he remembers today. Then the moment came:

Andrew stepped into the room, the 'elevator doors' closed, and he felt effects not unlike the rapid special displacement of a fast moving elevator. About ten minutes later, the doors opened and he found himself in a large underground concrete bunker - apparently a facility beneath the surface of the planet Mars.

Andrew noticed that there were two round openings near the roof of the bunker and by some means he was able to climb up to them. He did so and stepped through one of the holes to find himself on the surface of the Red Planet!

Andrew explained that he had not been equipped with a spacesuit - nor did he need one - although he had been provided with a bubble-like helmet to wear with a supply of oxygen attached to it. Otherwise he was dressed in nothing more than jeans, a T-shirt and tennis shoes. Stepping out onto the surface of Mars, he encountered a warm, windy climate. Also, he immediately saw two other people standing there, 'as if they were just a couple of people hanging around a food vendor on any city street.' He noticed that these two individuals were not wearing spacesuits or helmets. He stepped up to them and asked them if he really needed his bubble helmet. They told him it was not necessary.

Andrew removed his helmet and said he encountered a Martian atmosphere that was breathable, although the air felt 'hot and smoggy' and was somewhat difficult to breathe.

Here again Andrew's tale turns vague. He said he had a desultory conversation with the two individuals - whom he said were not 'Martians' but normal human beings - after which he wanted nothing more than to return to the teleportation device and get back to earth. Andrew reported that he then clambered back down through one of the holes he had emerged from, re-entered the teleportation device and, after waiting about a half hour, he was transported back to earth.

This was to be the first of a number of journeys he would make to the Red Planet. He concluded, astounding as it may seem, that the US Government had been engaged in a *Mars Project* for a number of years by the time he made his first visit there in the early 1980s. He said that there may be as many as 600,000 human beings living and working on Mars - all top secret and hidden from the general American and world public.

**What to Make of Andrew Basiago's Stories**

We have only touched on the most basic details of Andrew Basiago's extremely variegated stories to have been a top secret government time traveller, and to have made a number of routine trips to Mars. There are literally hundreds of hours of recorded conversations with Andrew available on numerous Internet sites, including radio interviews, video interviews and prime-time television shows featuring his story.

He says that all of his experiences had been erased from his memory, but that he somehow began to spontaneously recall everything that had happened to him in the late 1990s. This launched him on an investigation of Project Pegasus, using his skills as an attorney and researcher to rebuild his forgotten past, and to verify and corroborate all that had happened to him during the early years of his life.

Andrew says that he is now 'the first whistle-blower' and is determined to pry the lid off of what he calls the most massive government cover-up of all time. He says that people have a right not only to know, but also to enjoy the advantages of teleportation technology and the considerable economic improvement it could bring to our society.

If nothing else, Andrew Basiago is a spell-binding storyteller. He is able to speak for hours at a time, weaving complex tales of time travel and teleportation to Mars without contradicting himself, or exposing obvious holes in his story. In one Internet-radio interview, for example, he speaks for six uninterrupted hours, telling tale after astounding tale of his former life as a secret time traveller for the US Department of Defense and as an interplanetary teleporter for the CIA.

A television interviewer at KOMO4 in Seattle described him as a 'rapid-fire spontaneous weaver of science fiction stories' except that Andrew maintains his story is fact and not fiction. He also claims others have come forward to corroborate his claims, including several of the original one hundred and forty children that had been enrolled in the time travel programme with him in the late 1960s.

Sceptics dismiss Andrew as the worst of crackpots, of course, but his critics have not slowed him - not a bit. He maintains a web page dedicated to his Project Pegasus experiences, and another about what he calls his on-going research about 'life on Mars'. He also plans to release a book about his experiences, but says his busy private law practice affords him little time to get the book completed.

Whatever the case, Andrew Basiago has become a staple for Internet forum debates and podcast interviews. He has garnered attention from many sources in the mainstream media, including high-profile television shows which focus on conspiracy theories and paranormal topics. He certainly has captured the imagination of a certain core of believers - and as long as there are people who will listen to him, Andrew Basiago is likely to continue to tell his story.

# Planes, Trains & Automobiles ...And Ships

**Planes**

It was 1974 and a National Airlines 727 requested landing instructions from the control tower at the Miami airport. The captain was surprised when he was instructed by a panicked sounding flight controller to put his plane down on an isolated runway reserved for security risks and special situations, such as when planes are hijacked or in trouble.

When the jetliner landed without incident the captain, crew and entire compliment of passengers were surprised when security forces rushed onto the plane and hustled everyone off board, sending them through a special disembarking routine.

Emergency vehicles, fire trucks and airport police were on hand outside the aircraft.

Everything about the flight had been 100% routine, so captain and crew were mystified about what all the excitement was about. After everyone had calmed down, the captain finally demanded to know what was going on.

The head of airport security told him: 'You guys disappeared into thin air for 10 minutes. You might want to take a look at your clock.'

He was referring to the jet's on-board chronometer which showed that it read 9.20am. The captain glanced at his wrist watch - but it matched the clock on his plane's control pane. 'Looks alright to me,' he told the security chief. Then the chief showed him his watch. It read 9.30am.

As reported in the Miami Herald, an investigation showed not only the onboard clock of the 727 had lost 10 minutes - but so had the personal timepieces of every passenger on board!

**Lost Radar Contact**

The panic at the Miami airport started when air traffic controllers were looking right at the radar screen blip of the 727, and saw it vanish from the screen. They immediately assumed that the plane had dropped precipitously to the ground in a crash landing, or perhaps exploded in mid-air. The radar unit was functioning normally and was still registering other flights in the area.

Helicopters and rescue units were sent to a swampy area west of Miami where they were certain the National Airlines plane had gone down. For ten frantic minutes, no one was able to find anything.

Then, almost like magic, the jet reappeared on the radar screen and in the exact same location where it had been ten minutes previously. During those ten minutes, the blip should have moved considerably from one position on the screen to another. Furthermore, several other planes passed through the area where the vanished plane had been.

From the point of view of pilot, crew and passengers, absolutely nothing had happened. If they lost ten minutes it had gone completely unnoticed by anyone. The pilot and co-pilot said they experienced no break in communications with the control tower. In short, they experienced absolutely nothing abnormal. But the fact remains the National Airlines flight did not exist for a period of ten minutes. It was more than a matter of a radar malfunction - after all, where was the plane? It should have landed ten minutes before it did. If it had stayed in the air and strayed off course, the pilots would certainly have been aware of it.

The case of the National Airlines 727 has never been solved. It has been written about in a number of books. Many suggest there is only one logical explanation for the event based on the observed facts: The airplane winked out of existence for ten minutes, did not exist, but then reappeared. It somehow skipped across ten minutes of time.

Did it enter some kind of space-time anomaly in the air? Was there a wrinkle in the fabric of reality? The true answer may never be known.

## Trains

Russian Prince Felix Felixovich Yussupov will forever be remembered as the man who shot Gregori Rasputin in the back. Felixovich was one among three members of Russia's pre-revolution royal elites who had had enough of the 'Mad Monk' and his meddling in the affairs of the last Tsar, Nicholas II, and his wife, the Tsarina Alexandra. Felixovich and his cohorts first poisoned Rasputin, shot him multiple times, beat him with an iron bar, stabbed him and threw him into an icy river - and still he crawled back out.

Yussupov and his conspirators eventually got the job done, but Rasputin's seemingly inability to die would not be the only paranormal event in the life of the Russian Prince. One such event involves an encounter with a ghostly train in the deep forests outside of Moscow, and what seems like a classic case of a strange time-shifting anomaly.

In his book, *Lost Splendor*, Prince Yussupov recounts a remarkable story which he dubbed 'a supernatural event'. Here is an excerpt from Chapter 8:

*'One year, toward the end of the holidays, my brother and I had a strange experience, the mystery of which was never solved. We were leaving by the sleigh from Moscow to St Petersburg. After dinner we said goodbye to our parents and entered the sleigh which was to take us to Moscow. Our road led through a forest called the Silver Forest which stretched for miles without a single dwelling or sign of human life. It was a clear, lovely moonlight night. Suddenly in the heart of the forest, the horses reared, and to our stupefaction we saw a train pass silently between the trees. The coaches were brilliantly lit and we could distinguish the people seated in them. Our servants crossed themselves, and one of them exclaimed under his breath: 'The powers of evil!' Nicholas and I were dumbfounded; no railroad*

*crossed the forest and yet we had all seen the mysterious train glide by.'*

Felixovich Yussupov was an adolescent when he saw the ghostly train, which was in a vast area known as the Silver Forest. Around the year 1900 when this event occurred, there were no rail lines running through this area. Also the reaction of the peasant servants would indicate that they were witnessing something extraordinary, and something which should not have been there.

Interestingly, a rail line would later be built through this exact area of the Silver Forest, but not until 1925. Did the young Russian Prince, his companion and servants encounter a brief crack or window through the fabric of time, seeing a train twenty five years into the future?

**Automobiles**

The accounts of people encountering shifts in time while travelling in cars are incredibly numerous. They range from brief encounters with eerie out-of-place hitchhikers dressed in the clothing of another era, to incidents of driving through mists and fogs that act like time-altering doorways.

Some have attributed this to something called *highway hypnosis*, also sometimes called *white line fever*. The theory is that when a motorist has been driving a long distance, the apparent motion of the dotted centreline produces a hypnotic, or trance effect on the driver. The highway hypnosis phenomenon was first published in a scientific peer reviewed journal of psychology in 1921 and has since been studied a number of times.

Interestingly, one of the symptoms attributed to highway hypnosis is amnesia. Psychologists speculate that drivers enter into a deep trance state, or a condition called *sleeping with your eyes open*, and thus tend to forget a certain portion of their journey. When they stop, the person is surprised to be miles away from where they thought they were, and they have the feeling they are 'missing time'.

Still others point out that such a condition is extremely rare, while some psychologists dispute that people could actually drive safely for any significant distance in a trance state. The most likely result is a crash.

Highway hypnosis also cannot explain cases like the following which was reported by the Winnipeg Free Press of Winnipeg, Manitoba, in Canada.

Charles Dunay and his friend Donella Stronk were making the long drive across the vast flat plains of Manitoba prairie country. They were headed for The Pas in north central Manitoba, an all-day jaunt. The drive along highways in this region are monotonous because the roads seem to extend off into an unending flat table top where there are few cities or landscape features, such as trees, to break up the scenery.

As they motored along in their Dodge SUV, the couple noticed a fast travelling black Chevrolet sedan pull close up behind them. Charles was a bit miffed because the Chevy tail-gaited him for a period before passing. When the car finally pulled out to pass,

Charles and Donella were mildly amused by the Chevrolet's vanity Minnesota license plates, which read, *Wonky*.

The car quickly sped on ahead and was lost from sight. Charles and Donella continued to drive. About ten minutes down the road, they spotted a roadside vegetable stand. A large sign announced that the stand was offering fresh organically grown vegetables sourced by a local 'Nature Commune'. The couple considered stopping, slowed, but then decided they needed to keep going to make good time on their way to The Pas.

About forty miles down the highway, another car appeared in the rearview mirror - another fast-driving tail-gaiter. This car was also a black Chevy, and sure enough, it was adorned with the Minnesota licence plate, *Wonky*.

Naturally, Charles and Donella assumed that *Wonky* had stopped at the vegetable stand, even though they didn't notice a black Chevy parked there as they slowed and then drove by. Once again *Wonky* tailgated for a bit, then pulled out to pass and sped off into the distance.

Another ten minutes down the road, Charles and Donella saw a sight that made them uneasy - there was the vegetable stand again. It was the same stand with the same sign advertising organic vegetables sold by a 'Nature Commune'.

Could it have been a second vegetable stand? 'No way,' said Charles, and Donella agreed.

The couple were intrigued and so, just to be sure, they decided to stop at this 'second' vegetable stand and ask the proprietors if there was an identical vegetable seller about forty miles down the highway. A woman working there said this was the only such stand in the region, and that it sold vegetables produced by a Nature Commune just three miles from the roadside location.

Charles said: 'I've thought about it a thousand times since then. The only conclusion I can come to is that we experienced some kind of slip in time or some sort of dislocation in space. The black car passed us twice, and we saw the vegetable stand twice. It just doesn't make any sense. Since Donella had the same experience as me, there was no way that both of us imagined what happened.'

**Ships**

Any discussion of time slips relating to ship maritime incidents inevitably leads to the hundreds of stories that come out of the famed Bermuda Triangle, an area of the Caribbean Ocean with the city of Miami, Bermuda and Puerto Rico at the points of the triangle.

The legend of the Bermuda Triangle began in 1950 when Associated Press reporter Edward Van Winkle Jones published a story about the unusual and unexplainable disappearance of a ship in that region. It wasn't until the 1960s however that the term Bermuda Triangle entered into the mind-set of popular culture.

The majority of the stories concerning incidents within the Bermuda Triangle involve vanished vessels, or aeroplanes, but many others involve only the disappearance of the crew and passengers, with the ship left intact and in perfect condition.

However, long before strange happenings began to be reported from the Bermuda Triangle, a number of ships have returned from sea with tales of astonishingly odd incidents, many of which resemble the *time storms* we introduced you to earlier.

A case in point is a story reported in the New York Herald dated 1904. The British steamship *Mohican* piloted by a certain Captain Urquhart was sailing along the east coast of the United States when it suddenly encountered a *thick luminous fog which had the immediate effect of magnetizing everything on board.*

The compass of the *Mohican* began to spin wildly out of control - spinning as fast as an electrical engine. The ship had a metal deck and anything else made of metal was glued to the floor. Tools, heavy anchor chains and anything else was stuck so hard that none of it could be budged.

The grey vapour which surrounded the ship was said to glow like phosphorous. It also produced 'glowing, flashing lights'. At one point, the entire vessel looked as if it was on fire and the sailors took on an appearance of 'glowing phantoms'.

The Captain is quoted in the article:

*'The seamen were in terror. Their hair stood straight on end, not from fright so much as from the magnetic power of the cloud. They rushed about the*

51

*deck in consternation and the more they rushed about the more excited they became. I tried to calm them, but the situation was beyond me.*

*'For a half-hour we were enveloped in that mysterious vapour. Suddenly the cloud began to lift. The phosphorescent glow of the ship began to fade. It gradually died away and in a few minutes the cloud passed and we saw it moving off to sea.'*

The ship was rendered completely motionless during its time in the glowing fog. The crew reported an 'eerie feeling' as if they had 'left the world' and that 'time did not exist, or had completely stopped'.

The *Mohican* incident bears all the hallmarks of a classic roving *time storm* scenario. Some have suggested that similar incidents can explain the seemingly unusual number of vanished ships or disappearing crews and passengers of ships sailing the Bermuda Triangle - but unlike the frightened crew of the *Mohican* - they don't stick around to talk about it. They're gone.

But where do they go? Are people sucked through time portals never to return? But how can just the people of a ship vanish while the vessel remains in place? Are entire ships sailing through inter-dimensional time doorways placing them in some far off ancient sea or the waves of a future ocean?

Most scientists reject such supernatural reasoning and contend that lost ships and crews have wholly natural explanations - such as pirates, freak storms, rogue waves - yet they have little to say about events such as those described by the crew of the *Mohican*.

For the captain and entire crew to all agree that the strange phenomenon occurred as they said it did makes for a compelling case that scientists don't have all the answers - and the floating clouds or *time storms* may be a regular, albeit rare, occurrence on the high seas.

# Liverpool

The local media of Liverpool in the UK has been unable to ignore a number of highly peculiar stories over the past five decades or so. Whether the hard-nosed journalists of the Liverpool Echo and the Liverpool Post care to admit it or not, people in their city and in the surrounding area seem to be experiencing regular slips, or dislocations, in time.

A 1974 story in the Liverpool Echo tells the story of two twenty something women, Sandra and Jill, who had decided to visit a discotheque which was located at that time in the Leighton Court Country Club in the nearby village of Neston.

Jill's brother Andy agreed to drive the girls to the disco. Once there, Sandra and Jill met up with another friend, Shelly.

The evening proceeded as one might expect at a busy disco packed with dancers, loud music provided by a DJ and a dance floor replete with dazzling light effects. At one point, Sandra spotted a man she knew and on whom she had a crush because he bore an amazing resemblance to a popular British TV actor of the day.

Sandra told her friends she was going to try to 'dance her way' through the crowd of other dancers so she could get near her crush. Jill and Shelly decided to sit back and watch their friend as she attempted to 'make her conquest' - only to be frightened when

Sandra appeared to literally vanish from the middle of the dance floor.

At first they thought it might be an effect of the disco lights and noise in the crowded, smoky disco environment. On the other hand, they had their friend in direct sight when she simply 'winked out'.

'It wasn't like she was just suddenly lost in the crowd,' one of the girls told a Liverpool Echo reporter. 'It was like one minute she was there, and the next minute she was gone.'

Both Jill and Shelly became alarmed and searched the dance floor for Sandra. She was nowhere to be found. Next they looked in the bathrooms and all other areas of the discotheque. It was as if Sandra had vanished into thin air.

Fearing the worst, they were about to call the police, when suddenly, Sandra seemed to reappear near the spot where they had last seen her. She also seemed dazed and confused as she made her way out of the dancing throng back to her friends. And she had a remarkable story to tell.

Sandra said that as she elbowed her way through the dancers, the volume of the noise in the disco suddenly doubled to an incredibly loud racket. It still sounded like music, but this was music of a much different kind. 'It was more like a harsh, angry, throbbing noise,' she said.

Even stranger than the harsh music, however, were the bizarre mob of dancers she now found herself surrounded by. Unlike the polyester prints and glam suits of the 70s disco era, this crowd were dressed like extreme punk rockers - if they were dressed at all.

Sandra said that many of the dancers, both men and women, were naked and most of them were covered with tattoos and various body paints. The flashing sparkle of a spinning disco globe had been replaced by garish laser beams stabbing and darting throughout the crowded interior. Some of the dancers wore bizarre masks with long noses. She remembered one man in particular - he was naked, his hair was dyed a bright gold and his nails were painted a vibrant yellow. This man approached Sandra and attempted to tear her clothes off. Sandra fended him off with a slap.

Just as she began to panic, Sandra suddenly found herself back in 1974 among disco dancers - which seemed incredibly tame compared to the wild rave she had just experienced.

Since this story appeared in 1974, long before the 'new wave' of punk rock transformed the music and nightclub scene, the suggestion is strong that Sandra somehow leapfrogged into the future for a few minutes - into a future in which the punk styles and music had become even more extreme. It is unlikely that a twenty-four year old British girl could have anticipated and so correctly described a punk-like social trend that was still at least six years into her future.

## A Liverpool Cop Slips Back to the 1950's
The Liverpool Echo reported another time slip story in 2003, this time involving a man, who was an off-

duty police officer, and his wife on a shopping excursion on Liverpool's Bold Street. This story was submitted by a correspondent for The Echo, Tom Slemen.

Upon arriving in the Bold Street area, Carol told Frank, her husband, that she needed to visit a local bookstore to buy a novel she had long wanted to read. At the same time, her husband spotted a friend and said he would stay behind and have a chat while she went to the bookstore. He then planned to visit a record store and said he would meet her back at the book store.

After taking leave of his friend, Frank went to the record store on another street and then made his way back to the bookstore, a location he was familiar with, and noticed it was adorned with a sign that identified the shop as *Cripps* - not the name of the bookstore, as he remembered it. It was supposed to be Dillon's Bookshop.

Just as he stepped out on the street to cross over to the store, he was startled by the beep of what he said sounded like an old fashioned motor horn. Sure enough, a fast-moving 40s or 50s model box van sped by. He noticed a name painted on the side of the van - *Caplan's.*

Suddenly, everything seemed odd.

He crossed over to the shop called *Cripps* and found not a bookstore where a bookstore should be, but instead a ladies shoes and handbag store. He turned around as he became aware that everyone was dressed in the styles of late 1940s or early 1950s.

Frank quickly began to feel panicky. Just then, however, he spied a young woman of about twenty-five years old wearing what looked like normal modern clothing. As he focused on her, everything appeared to return to normal. He turned around and saw that Dillon's Bookstore was back. The young woman had walked into the bookstore, but quickly turned and had come back out.

Frank tapped her on the arm and asked her, 'Did you see that, I mean, did you see anything strange just now?'

She replied with a nervous laugh: 'Why yes, I thought I saw a new shop and went in to take a look, and discovered it was a bookstore. I thought it was something else.' Then she walked away.

Reporter Tom Slemen was later invited to speak about the case on a local radio station. Upon hearing his tale, several people called in to say that they remembered when there was a shop in that location on Bold Street called *Cripps*. Others also remembered a firm called *Caplan's*, the name on the side of the van.

Frank's wife, Carol, said she had noticed nothing unusual. She bought the book she was looking for, and could only listen with disbelief to her husband's strange tale.

### The Time Slip Discount
Yet another woman found herself in a Liverpool establishment that shouldn't have been there in 2011, yet she swears it was.

Seventeen year old Imogen, delighted that her older sister had just given birth to a baby girl, took a trip down to the Liverpool City Centre to purchase some gifts for her new niece.

She spied a store called *Mothercare* on the corner of Lord Street and Whitechapel and strolled inside where she was delighted to find a wide selection of adorable baby clothing items at remarkably low prices.

She picked out a variety of items, including some polka-dot bibs and a pink cardigan sweater. It seemed high quality stuff, yet the price was remarkably low. She naturally reasoned that since this was a newly opened store, it must be a grand opening sale of some sort. When she took her items to the counter, however, she was in for a confounding surprise.

The clerk rang up her items and Imogen handed over a credit card. The clerk looked at the piece of plastic with suspicion. She then excused herself and took the credit card over to the store manager. The manager squinted at the card and returned to the counter where she told Imogen: 'We don't take those, love.'

Imogen was a bit flummoxed but, on the other hand, having a credit card rejected was not enormously unusual. She had little cash with her, so she returned the items and left the store.

Back home, she told her mother of the unusual event at the *Mothercare* store. Imogen's mother told her daughter that she must be getting her facts mixed up, for the simple reason that there was no *Mother-*

*care* on the corner of Lord Street and Whitechapel. Her mother could be certain of this because she remembered when there actually was a *Mothercare* at that location in the early 1980s, but that it had been replaced by a bank years ago - and Imogen's mother could be certain of this because this was her own bank.

Imogen was in total disbelief. She was certain beyond a reasonable doubt that she had been in the *Mothercare* store at that specific location - she even urged her mother to go with her the next day back to Lord Street to prove she was right. They did so and, sure enough, there on the corner of Lord Street and Whitechapel was a bank.

Upon reflection, Imogen began to recall that some of the details of her shopping excursion to the *Mothercare* store were peculiar, mostly the style and dress of the employees. They were not outrageously out of place, yet the hairstyles and dress seemed dated. It didn't seem terribly unusual to her at the time, but now these details took on a new significance.

Imogen remained solidly convinced that she had been in the *Mothercare* store on Lord Street in 2011, even though there was no way it could have been there.

**Future Cars and More**

Stories of time slips in Liverpool - most of which seem to be centered in the Bold Street area - are almost too numerous to mention. Most of them take on a similar tack to those you have read about here - or-

61

dinary people going about their days, shopping or walking to work, or stepping out for lunch - only to step into another decade in time.

Sceptics suggest that after the first couple of original time-shifting anecdotes were issued by local media and became widely talked about, an urban legend developed around them, or perhaps a kind of local mythology. One story spawned another, and before you knew it, people began to interpret even minor experiences of the unusual as slips in time.

On the other hand, literally dozens of people have reported experiences that have no good explanation, and these experiences are often corroborated by neutral, third-party witnesses.

One particular sighting in Liverpool that has become almost legendary is that of the *futuristic car*. It's been seen by dozens of people.

The first sighting of the futuristic car dates way back to 1957 when a businessman by the name of George Kingsley was motoring along the Queensway Tunnel. It was late at night, near midnight, and the tunnel was all but deserted when George saw in his side-view mirror an incredible sight.

It was a sleek, futuristic bright yellow-gold vehicle - triangular shaped with aerodynamic rounded edges, travelling at an extremely high speed. The rakish 'future car' went screaming past George at a lightning pace. Up ahead the tunnel curved and the golden car went careering right into the wall. George was expecting a frightful crash and explosion but saw something even more astonishing - the vehicle appeared to pass effortlessly right through the wall of the tunnel! It was gone!

And yet the 'future car' left behind skid marks on the highway in the spot where it seemed to veer out of control. What caused the crash? Did the driver in the distant future suddenly find himself transported into 1950s England, resulting in panic and loss of control of his vehicle?

Since then, more than eighty people have reported sighting the amazing futuristic 'golden car', catching just brief glimpses of it in and around the environs of Liverpool. Critics like to cry, 'Mass hysteria!' Maybe, but such a theory does not account for all of the other circumstantial evidence of other reports of time slips in similar, everyday situations.

Why would ordinary people, such as police officers, mothers, businessmen and sundry citizens from all walks of life insist they experienced spontaneous visions or brief transportations to alternate-time environments? Is the city of Liverpool located on some kind of confluence of natural energies which cause the veil between dimensions to wear thin, or occasionally break down?

The speculation might be endless, but one thing is for sure: Dozens of people over the past several decades are convinced they became briefly displaced in time - and you'll never convince them they didn't.

# Back in Time – Retro-causation

Can you send a message back in time to yourself?

For example, let's say you're a college student and you have a big final exam coming up. Wouldn't it be great if you could leap ahead in time, get all the correct answers to the test, and then time travel those answers back to yourself so that you could ace the test?

Believe it or not, this is not only possible, but has proven to work in a number of scientific and peer-reviewed laboratory studies.

It has been established beyond statistical doubt that the effects of time do indeed flow backwards and forwards, and that we may be able to access that two-way stream of time to help ourselves in tough situations that have happened to us in the past, and which affect our present day lives.

Sounds too bizarre to be true? First consider this statement by physicist Nick Herbert, author of the book, *Quantum Reality*:

*'If time is like space, then the past must literally still exist 'back there' as surely as Moscow still exists even after I left it. If the past still exists, then it makes sense to consider whether one could actually travel there ...'*

He continued:

*'If we take the fourth dimension seriously, we must believe that past and future have always existed, and that human consciousness, for reasons we do not comprehend, perceives this 'block universe' one moment at a time, giving rise to the illusion of a continually changing present.'*

## The Reverse-Time Psychology Experiments

That the past still exists, and that we can also send messages or information into the past, has been proven by one of the world's most respected psychologists, Daryl Bem of Cornell University in New York. His experiments were first reported in the Cornell Daily Sun and then quickly received worldwide media attention.

Here is how Daryl's experiment was set up:

He enrolled into his study one hundred volunteers from among Cornell students, fifty males and fifty females. The students were brought to a room and seated in front of a wall on which were hung two curtains.

Behind one curtain was just a blank wall, but behind the other was an erotic picture featuring an exciting and stimulating sexual situation. The students were asked to click on a button which would select the curtain behind which an erotic picture was hiding. All they knew was that one side was blank and the other had some kind of picture.

The result was that almost 54% of the students always chose the curtain containing the steamy picture, while 46% chose the blank wall.

66

This may not seem like a highly significant difference beyond the 50-50 that pure chance would predict; however, the likelihood that the 54% success rate could have happened by chance were calculated by physicist Edward Close to be 1 in 74 billion.

Daryl conducted this experiment and others like it for more than eight years, and his results were always consistent. All of his volunteers were somehow 'reading the future'.

More experiments by Daryl, and then by other researchers followed. It was found that people from all walks of life could significantly improve scores on tests - university exams, civil service tests, motor vehicle license tests - if they studied for the test before - and then after taking the test.

For example, let's say you need to take your written exam to get your driver's licence. Your test is on Friday. So on Thursday you study the driver's manual and take the test the next day, Friday. Then, on Saturday, you study the manual again.

This seems ridiculous from a normal way of thinking. If you passed the test on Friday, it would seem like an absolute waste of time to take a few hours to study the test you have already completed! But the fact is, if you do make that commitment to study the day after you take the test, you will have a chance of passing your test that will be significantly statistically higher.

Daryl Bem and other researchers reached this conclusion: It meant that our *future selves* are sending information back in time to help out our *past selves*.

In effect, these people were enabling a form of actual time travel.

But wait a minute …

Wouldn't the effect of the Bem 'erotic picture experiment' and other similar experiments better be explained more simply by ESP or precognition? Why bring in the element of time travel at all? Well, when you think about it, the idea of ESP or precognition is already halfway there. Those who claim to have the ability to see into the future are already slipping beyond present time with some aspect of their cognitive ability to see another time existent in the future.

As physicists and theoreticians began to develop concepts of time in terms of the structure and dimensionality of time, it was inevitable that some would attempt to fit ESP and precognition into that framework. From where we are positioned in time right now - the present - we can 'look back to the past' with our memories. But we might also be able to send messages down through the corridors of time to affect real change.

### Retro-Causality

Scientists call this retro-causality - when some action you can take in the present can actually alter events in the past, and thus, reshape your entire timeline.

There have been a number of experiments which have attempted to show that sending a physical object back in time is actually possible. To this end, physicists have attempted to send a single particle, such as a photon, into the past, and do it in a way that is verifiable and provable.

Although highly controversial and the debate is hot and ongoing, some scientists claim to have demonstrated that they can send tiny quantum particles back in time to interact with themselves. A study done by a team of MIT (Massachusetts Institute of Technology) scientists published in May of 2012 said:

*'Closed time-like curves (CTCs) are trajectories in space-time that effectively travel backwards in time: a test particle following a CTC can in principle interact with its former self in the past. CTCs appear in many solutions of Einstein's field equations and any future quantum version of general relativity will have to reconcile them with the requirements of quantum mechanics and of quantum field theory.'*

## The Possibilities

While sending a tiny subatomic particle into the past doesn't seem like a significant accomplishment, the fact that it might be possible throws open the doorway to the prospect of time travel into the past, or the future. If a single particle can travel into the past, and interact with the past, that means anything can. It's just a matter of scale and degree.

Many people have already jumped in with both feet to make much bolder claims. They say that we can actually change the past to improve our lives today. They say that's why certain modern forms of psychological therapy work.

Psychologists say that if we 'reshape our traumatic memories' we can gain healing in the present. For example, let's say that an adult woman is depressed because she grew up with a cold, cruel or uncaring father. Through a process of deep meditation, possibly hypnotherapy or other methods, the woman could 're-imagine' what her relationship with her father was like. She could strive to 'replace' memories which have haunted her for her entire life with 'made-up' memories of happy, caring times with her father.

This kind of therapy has helped millions of people cope with the past and make themselves feel better today. Psychologists say that if the re-imagined events are not real, it does not matter, because the healing effect is the same as if the past literally was changed.

But some - mostly philosophers, metaphysicists and New Age thinkers - have suggested that when we strive to reshape our memories, we are doing much more than that - that we are actually physically, literally reshaping the past. We are changing time. We are enabling the power of retro-causality to alter what really happened, and thus are changing the entire timelines and our very realities.

Among those who have suggested this is the highly respected Dr Deepak Chopra who co-authored a number of articles about the retro-causality effect with computational physicist Menas Kafatos, PhD, of Chapman University.

Also, more than a few popular writers have already claimed to have the ability to reshape their present day realities through the power of retro-causality. One is Richard Bach whose numerous books, including *Jonathan Livingston Seagull* and *Illusions,* sold multi-millions of copies. In his book, *Running From Safety*, Bach describes how he 'contacted' his boyhood self and engaged in an extended conversation with this past self - which resulted in a tremendous amount of mental healing for his present state of mind.

Richard said that his book 'cannot be considered fiction' and that his interaction with his past self has a kind of reality that was something far greater than merely an extended effort to come to grips with memories of the past. Richard feels he may have reshaped his entire timeline.

Again, if this seems just too far-fetched for you, reconsider those experiments by Professor Daryl Bem of Cornell University. His results show clearly that by studying for a test AFTER you take the exam, you can significantly improve your score - even though you already took the test in the past.

That we can reshape our own pasts through a form of time travel has astounding implications. Better yet, the possibility that this is a real option has at least been partially demonstrated with physical experiments in the lab. Many have already claimed to have obtained healing and a 'remodelled' and better present by tunnelling through to their own pasts to change things for the better.

Could it work for you?

# The Kersey Time Slip

Was it a slip in time?  Or maybe it was a case of 'ret-ro-cognition'?  Either way, the bizarre experience of three young British Navy Cadets in 1957 would haunt them for the rest of their lives.

Their case would attract the attention of multiple newspapers, and was even written about in the stodgy and highly respected American periodical, *Smithsonian Magazine*.  The case also was prominently featured in a book by Scottish writer Andrew Mac-Kenzie.

The events took place in the Babergh District of Suffolk in the east of England.  Three young men, all aged fifteen, had recently signed up to join the British Royal Navy.  As part of their early training, they were set with a map-reading and coordinates-finding task. They were to navigate on foot across about five miles of the English countryside.  As it turned out, their target location was the tiny village of Kersey in Suffolk.

None of the three boys were familiar with the area. William Laing was from Perthshire, Scotland, Michael Crowley from Worcestershire and Ray Baker was a native of East London.  Nevertheless, they had successfully followed the map coordinates given to find where they were supposed to be - the village of Kersey.

Upon approaching Kersey, an almost overwhelming feeling of strangeness came upon the three adventurers, and it wasn't a good sensation. The boys said an oppressive mood invaded their minds. All three agreed that a penetrating sense of depression and subtle fear had inexplicably descended upon them. They also described it as 'a great sadness'. What's worse, they had the tingling feeling they were being watched by people who were hiding, people who were both fearful and unfriendly.

This is odd considering they were approaching a lovely, peaceful and picturesque village known for its delightful Old World charm and welcoming attitude toward visitors. But on this day, the streets of Kersey seemed deserted. There was also no sign or movement of anything modern - no automobiles or even bicycles. There was no drone of an aeroplane, no TV antennas on roofs, and they could see no telephone poles, wires, or streetlights.

Attempting to marshal their feelings of gloom and dread, the boys sauntered into the village and were struck by its shabby appearance. All of the houses seemed decrepit and crude. In fact, many were in a state of disrepair. One of the youths later described them as looking, 'hand built of rough-hewn lumber and timber framed.' The village was medieval in appearance.

The strangeness of the village caused the boys to recheck their coordinates. Perhaps they had wandered into some kind of long-abandoned dwelling area? They doubted that they had actually found Kersey, but they concluded that they simply must be in the right place.

The boys approached one of the nearest buildings which they said was fitted with smallish grimy windows. They pressed their faces to the glass and saw inside what looked like a filthy butcher's shop. They saw the skinned and butchered carcasses of oxen and found the appearance of the partially carved animals to be revolting. These carcasses were green with mould. There was no other furniture inside the butcher shop, and no people were evident. It was as if the animals had been partially butchered, then abandoned.

They also recalled other aspects of the 'butcher shop' structure - it had a crudely painted green door and small windows that were coated with grime and grease. They could not imagine that health inspection authorities would ever allow conditions such as these in any location processing meat or food for general consumption.

They moved on to another house which also had small windows and which was also caked with a film of muck. Inside they saw crudely white-washed walls, and again no furniture. The rooms were cramped and not of modern design. They began to get the feeling they were in some kind of 'ghost town' - and even though they encountered no people, they felt certain they were being watched. They felt that, for some reason, everyone was hiding from them, as if they feared them. By now they were thoroughly spooked. They turned and hurried their way out of the eerie village. They climbed a small hill and did not turn around until they reached the summit which was a considerable distance from the 'haunted village'. But when they surveyed the village from their vantage point on the hill, the whole setting seemed to have changed. They noticed smoke coming from chimneys and they heard the peel of a church bell drifting out from the town. They thought they saw people moving about in the streets. In short, it looked like a normal small English country town.

All three had had enough, however. As a feeling of eerie dread was still upon them, they turned and ran. They said it took several hundred yards for their fear and depression to leave them - and they didn't want those feelings to return - so they kept going and returned to their base of operations. They told their supervising officer what they had experienced. He doubled checked their coordinates and confirmed that they had actually been in Kersey. He also 'laughed off' the boys' tales of a medieval village where a modern town of 1957 should have been.

In the years and decades that followed this 1957 experience, what they had gone through continued to trouble two of the three men, although one of them, Ray Baker, more or less forgot about it. But William Laing and Michael Crowley found they couldn't get away from what had happened to them. As it transpired, in the 1980s they both found themselves living in Australia. They contacted each other and began to exchange letters about what they had experienced that day. For William and Michael, the experience still resonated deeply - they wanted to dig deeper for a possible explanation.

To this end, they decided to contact the Society for Psychical Research in London. They were put in touch with one of its top researchers, Andrew Mac-Kenzie, who eventually made the Kersey time slip experience the lead story in his 1997 book, *Adventures In Time*.

In Andrew's account, other details of the experience are flushed out. For example, one of the things the boys remember is that as they walked into Kersey, the total environment seemed to change. For example, it was a brisk October day when they reached Kersey but as they walked into the village, they found the trees green and verdant; the grass was long and lush. It seemed more like a warm spring day.

But there was also a kind of oppressive stillness and silence. The air became absolutely still, and felt cloying and humid. They remembered passing by some ducks in a stream that seemed 'frozen into inactivity' - it was as if even the ducks were frightened by an uncanny alternation in the very fabric of space. All birdsong ceased, as did the buzz of insects or the rustling of leaves. William Laing is quoted as saying:

*'It was a ghost village, so to speak. It was almost as if we had walked back in time... I experienced an overwhelming feeling of sadness and depression in Kersey, but also a feeling of unfriendliness and unseen watchers which sent shivers up one's back... I wondered if we'd knocked at a door to ask a question*

*who might have answered it?  It doesn't bear thinking about.'*

The story of the boys at Kersey caused researchers at the Society for Psychical Research to dig into the history of the village.  They wanted to find out if this settlement's history extended back into medieval times, which indeed it does.  It is known that Kersey was established at least as of year 900 AD because mention is found of it in an old Anglo-Saxon will.

In 1990, investigators invited William and Michael to return to Kersey so that they could retrace the journey they made more than 30 years previously.  This they did.  Walking through the streets of Kersey, they quickly identified the house that seemed to be the butcher's shop which they had encountered years earlier.  It turned out this same residence - or at least the same site where the current residential home is located - was once a butcher's shop!

Investigators were able to confirm that the house was a butcher's shop at least as early as 1790, and that a building had existed on this exact spot at least as early as 1350.

One curious aspect of the village puzzled the men on their 1990s visit, however, and it seemed to contradict the idea that they had experienced a slip in time.  In easy view of the 'butcher's shop' is the ancient St Mary's church, known to date back to medieval times.  It is believed that construction began on St Mary's in the mid-1300s.  It is a well-known landmark of the town and region.  Why didn't the boys see it in 1957? It seems it would be impossible to miss its tall, square castle-like spire.

There is an explanation: Construction of the church was abruptly halted in 1348 because of the outbreak of the Black Death, which devastated much of Europe, including the village of Kersey. If the boys had time-slipped to a period between, say, 1350 and 1400, the uncompleted shell of half the church would have been hidden by trees and brush. By 1420, the plague had receded and Kersey was experiencing a period of prosperity based on the burgeoning wool trade. St Mary's was completed after that.

A period of prosperity would also explain a major point of contention brought up by sceptics who do not accept the time slip scenario. Sceptics say that it was highly unlikely that there would have been any windows at all in most medieval homes of the day. Windows were rare and expensive in those times and generally only the wealthy could afford them. However, the fact that Kersey had begun to experience a considerable economic boom starting in the early 1400s resulted in the emergence of a middle class - it is not unlikely that installing smallish windows on average homes would have been a popular trend of the day.

Others have suggested that the powerful feelings of depression, desolation and fear the boys sensed was due to the Black Death. This would have explained why they encountered few people, and why the butcher's shop and the other home they looked into were abandoned. The plague was known to carry people away abruptly and swiftly. For there to be oxen abandoned in mid-butcher might well be explained by the unrelenting swiftness of the Black Death.

But wait - sceptics have also claimed to find another hole in the story - the fact that there was a butcher's shop at all. If the boys truly had slipped back to the 1300s or 1400s, it was unlikely that a village of the size of Kersey would have had a butcher's shop. That's because meat was not a daily part of the diet of the average resident of that era. Meat was expensive and a luxury and generally only eaten on special occasions, such as feast days and holidays.

On the other hand, how likely is it that there would have been oxen being butchered in 1957? And how could all three of the boys have imagined seeing the same thing? A grim scene of mouldy, half-butchered dray animals?

Those who believe the lads truly experienced a momentary displacement in time consider the sceptic's denial of the idea of a butcher's shop in this location around year 1400 to be 'a desperate attempt to explain away the events'. Rather, supporters of the time slip scenario say it was likely that oxen were being butchered there, and that when considering the chaos brought about by the Black Death, sacrificing oxen for scarce food may have been a likely desperate measure.

Some people have suggested a compromise explanation - that of 'retro-cognition.' This is a term coined by Frederick Myers, a founder of Britain's Society of Psychical Research. It is defined as:

*Knowledge of a past event which could not have been learned or inferred by normal means.*

In effect, retro-cognition is the opposite of precognition, which is when a person has a psychic vision of future events. For some reason a person experiencing retro-cognition is receiving knowledge of a past event, and in a way that is not memory - but a vision of something that happened not only in the past, but possibly centuries, or even thousands of years ago.

Even so - classifying the Kersey event as a case of retro-cognition does little in the way of providing answers, or a satisfying explanation of what happened. A very problematic aspect of the retro-cognition theory is that it does not explain how all three of the boys each had a spontaneous retro-cognative vision at the same time!

Whatever the case, the experience of William Laing, Michael Crowley and Ray Baker would seem to bear all the hallmarks of classic time slip experience. William and Michael remain convinced more than fifty years after that strange day in 1957 that they travelled to the Middle Ages of England. The how and why of the strange experience may never be understood or explained.

# Remote Time

The human brain is still the most advanced computer on the planet.

As a species, however, we achieved higher form, self-reflective consciousness just a few thousands of years ago. The true potential of what our marvellous minds and brains are capable of may still be largely unknown and untapped ..….

… and that may include the ability to time travel.

In fact, some people may have already leveraged their brains for use as their own personal time machines.

Consider the experience of Sergeant Mel Riley, a former member of the elite corps of 'psychic spies' developed and employed by US Military Intelligence. Mel was among the first of a hand-picked, highly select group of men who developed what is today known as *remote viewing.*

**Spontaneous Time Travel**

When Mel Riley was a boy growing up in a poor neighbourhood of Racine, Wisconsin, he frequently escaped his dreary urban environment for long forays into the Wisconsin woods. There  he could spend hours, sometimes days, just existing free in the wild - sometimes even foraging natural forest foods and building crude shelters to stay warm at night. No department store tents or packed lunches for Mel Riley!

On one such excursion when he was a boy around twelve years of age, Riley was in the vicinity of a ploughed corn field. But this was also an area known to be rich in Native American artefacts. Many arrowhead and other implements had been found there, left behind by the Indians since the Stone Age.

Walking along, Mel suddenly detected the pungent smell of wood smoke and cooking food. He also heard voices. He looked around expecting to see some campers but was stunned to see that the field of ploughed black soil was now a grassy plains area - and it was strewn with Indian lodgings with smoke coming out of them - as if this had been an established settlement for years! The only problem was, none of this had been there just minutes before.

Mel said there were cooking fires near the lodges and native peoples going about their lives, as if this was another era in time - and he concluded that it must have been just that. He felt that he had somehow slipped back into time, but the experience lasted only for a few minutes. An instant later, the ploughed field returned and the Indians were gone.

Mel told freelance science writer Jim Schnabel:

*'Things just sort of dissolved and there was the field again. It definitely convinced me that we could access different times, different places. Once something like that opens up to you, you can say, I can travel in time, like a time machine. I can go anywhere I want.'*

At the time, however, Mel Riley didn't know how it happened, or how he did it. He was convinced that he had literally accessed another century in time, but that it was a spontaneous event. There was nothing in his current understanding as a boy of the early 1950s to explain something which he was nevertheless convinced had been very real.

**Enter Remote Viewing**

Mel would get his answers much later after he joined the US Army, where he was trained as an interpreter of photographic intelligence information. His job was to examine high-resolution photos gathered by spy planes. As it turned out, he was not only good at his job, but unusually so. His ability to spot obscure objects in photographs was uncanny. Mel Riley could see things in the images that no one else could and he did so with a high degree of accuracy.

In 1970, Mel's ability attracted the interest of a super-secret programme that had recently been launched by military intelligence, the now famous (or infamous) *remote viewing psychic spying program.*

In short: Remote viewing is a specific technique in which people with demonstrated psychic abilities were tasked with using their powers to *look into* remote locations - employing only their minds - so that they could identify what was there. In the case of the US Military, the top brass was keen to spy on the Soviet Union. After all, in the early 1970s, the Cold War was in full bloom between the United States and the Soviets, and each was willing to try just about anything to gain an advantage over the other.

The US Department of Defense and CIA knew that the Soviets were working diligently on their own psychic spying programme, so the US Military was eager to develop its own. The attitude was: If they have it, then we better have it to.

The *remote viewing spying program* proved to be both highly useful, but also vexingly problematic and troubling - which is why it was eventually abandoned after some three decades and millions of dollars spent on honing the technique of psychic spying.

Interestingly, one of the problems with remote viewing was an issue related to time travel.

Here was the problem: Sometimes when a remote viewer was asked to describe the details of a particular location, he pinpointed buildings and other objects that later turned out not to be located on that site. Most of the time, they just assumed that the remote viewer had simply gotten it wrong, that his psychic abilities were out of tune that day, or some other not well understood factor was clouding the picture.

But then it was discovered almost by accident that sometimes the reason a remote viewer gave an inaccurate description was that he had sent his mind to the correct location - but not to the correct time!

In one case during a training session, a remote viewer was given the target of a specific location in a California city. After he described the site as best he could, he and his trainers visited the location. The remote viewer was convinced that there would be a large church-like structure with a tall steeple or spire nearby, but there was none. Other buildings were just where he predicted, yet others seemed out of place or missing.

Still, the remote viewer could not believe he had gotten everything so wrong, and decided to dig deeper. He looked into the past history of that particular city and the location in question - and discovered that the church with the tall steeple he had sensed with his mind had indeed been there - but was torn down thirty years previously!

He researched the records of the location further and discovered that his predictions were a 100% match for a period in time about forty years in the past! It seems that when he projected his mind to a specific physical location on the earth, he had also unwittingly sent his mind about forty years into the past.

It seems that our reality in time and space is a bit more slippery than we assumed. The time-anomaly implications which fell out of this realization were not lost on researchers. What it suggests is that the mind is not only capable of seeing things far away, but also capable of seeing things as they were far away in the past and future. The trouble is - how do you control the time coordinates? There seems to be no easy answer.

**Time Travel Experiments**

The slipperiness of time was just one factor which led to the government's abandonment of remote viewing as a viable intelligence gathering tool. The situation was actually more complex - however we will leave that story for another time.

But after the elite corps of *military grade remote viewers* was disbanded, many of them soon re-entered civilian life, and continued to pursue their work in remote viewing. Some of them wrote books, some went on lecture tours and others hired themselves out to private security firms or plied their trade for corporate clients.

Still others dedicated themselves to pure research. One of these men was Fred 'Skip' Atwater. It was Skip Atwater who spearheaded the remote viewing programme for US Army Intelligence. Upon leaving the service, he was hired as the Director of Research of the Monroe Institute in Faber, Virginia, a non-profit organization dedicated to the exploration of expanded consciousness.

At the Monroe Institute, Skip invited one of his best military remote viewers, a large beefy man by the name of Joe McMoneagle, to participate in a series of experiments that combined remote viewing and something called binaural beat brain enhancement. This involves listening to certain sounds which, in theory, help synchronize the two hemispheres of the brain to work in greater cooperation. Listening to binaural beat tones can cause a variety of effects, from sending subjects into deep trance or meditative states, to inducing out-of-body experiences.

It is one particular experiment Skip Atwater set up with Joe McMoneagle that interests us here because it involves a bold experiment in time travel. This experiment is described in Skip's excellent book, *Captain of My Ship, Master of My Soul.*

## Time Travel to the Planet Mars

In a normal remote viewing session the remote viewer is given a target location, although he is not allowed to see what it is. The administrator of the experiment writes down a specific set of geographical coordinates and seals them in a manila envelope. The remote viewer - who never gets to look at the coordinates written inside the envelope - then attempts to project his mind to that location to see what might be there.

In this case, Skip wrote down these instructions:
*The planet Mars, one million BC.*

He wrote it on a three-by-five index card, sealed it in a small opaque envelope, and asked Joe McMoneagle to put the envelope in his breast pocket.

In addition, Skip provided a specific set of coordinates which pinpointed a certain location on the surface of Mars. NASA probes had photographed this region, which revealed a number of anomalous looking structures on the surface of the Red Planet. Some have suggested they are of artificial origin.

Joe was placed in one of the Monroe Institute's isolated booths where he reclined on a comfortable bed with headphones that fed binaural beats into his ears. Joe achieved an extreme state of relaxation and trance, but was able also to stay aware enough to engage in conversation and take instructions from Skip who managed the experiment from a control booth.

Skip asked Joe to describe what he found as the recording tapes were rolling.

Joe first reported that he found himself in 'an arid climate' in some 'distant location'. (Again, he had no idea his target was a point on the surface of Mars.) He then said he was in a location that seemed to have suffered the 'after effects of a major geologic problem'.

He was then asked to move backwards in time - perhaps 10,000 years before the devastating geologic event. Joe was able to do so, and now reported that everything was totally different. He saw pyramids, and began to assume he was in ancient Egypt. He counted seven pyramids in total. He also reported that he was not alone in this location, but that there was the 'shadow of giant beings'.

Joe felt intuitively that the 'shadow' aspect meant that these giant beings were not there anymore, but once had been. It was as if he was sensing a trace of their former selves. Skip then directed Joe to travel back several thousands of years - deeper into time. He did so, and now reported that he was in the presence of 'a very large people wearing strange clothing'.

He described these people as being about twelve to thirteen feet tall with thin lanky legs and arms. He had an impression that they might be ancestors of the human race on earth, but that their civilization had been devastated by some kind of major geologic calamity, or possibly planet-wide disaster.

Joe reported an extended conversation with one of the beings. He describes this conversation in his book, *Mind Trek*. He eventually determined that he was on the planet Mars, and more than a billion years in the past.

But was he really there?

Critics and sceptics of course say: *'No way!'* They find such fantastic tales simply too much to swallow. On the other hand, it is difficult to explain how Joe correctly identified the pyramid-like structures that have been observed and photographed by Mars probes. The coordinates Joe was given (but which he never saw) were that of the location where the famous (and controversial) 'Face of Mars' is located and there are indeed pyramid-like formations near 'The Face'.

## The Future

If remote viewers can travel to the past with their minds, then they should also be able to 'travel' into the future - and many remote viewers have claimed to have done just that. One of the most significant remote view journeys to the future was accomplished by a colleague of Skip Atwater and Joe McMoneagle - the former US Army Major Ed Dames - also an elite member of that first remote viewer cadre of psychic spies.

Like others selected, Ed displayed superior abilities to identify targets with a high rate of success. After his service in the military, Ed became a private remote viewer contractor, of sorts. He also hit the lecture circuit and has written a number of books about remote viewing. He conducts remote viewing seminars and even offers an online study course.

Ed, in recent years, has spent a lot of time casting his mind into the future and he, along with other 'famous' remote viewers, are predicting dire climatic changes to the planet earth which will be produced by global warming. In short, what Ed and others predict sounds dire to the point of being cataclysmic.

**Accurate or Problematic?**

The problems with remote viewing time travel experiences are many, however. Firstly, there can never be a certainty that even the best remote viewers are always 100% accurate. In fact, remote viewing has never been said to have achieved perfect results - which is another reason among several that it was abandoned by US Intelligence as a regular part of its spy programme. The very best remote viewers are said to achieve accuracy rates of 75% to 80% - but even those levels were not achieved consistently.

Secondly, there is still much that is not understood about the time-space continuum, including the problem presented by the Multiple Universe scenario. Most physicists now accept that the MU scenario is probably the correct one. That means that when remote viewers send their minds into the past or the future, they may very well be also slipping into alternate universes - separate universes on different timelines from our own - and more importantly - with different sets of circumstances and outcomes than our own.

Thus, when Ed Dames peers into the future and sees a climatic disaster precipitated by global warming run amok, there is no way of knowing if he was viewing our universe and our direct timeline to the future - or one of the other infinite number of alternate universes predicted by the Multiple Universe scenario theory. He also may simply be getting it wrong. One of the biggest challenges for the remote viewer is weeding out the influence of his own subconscious mind, biases, fears and pre-conceived belief systems.

Whatever the case, the time travel implications of remote viewing are extremely fascinating. Men such as Mel Riley seemed to have experienced genuine visions of other locations in other centuries. If not all remote viewing time travel events can said to be 100% bona fide and real, it's almost certain that at least a percentage of them are real. The implications of that are astounding.

# Conclusion

It may be possible to pinpoint the modern day fascination with time travel to 1895, the year H.G. Wells published his ground-breaking novel, *The Time Machine*. As we have noted in this book, humankind's obsession with time is ancient, but it was the advent of modern technological devices that helped us frame the possibilities in a whole new way.

The vivid adventure of a British gentleman only identified as *The Time Traveller* in Wells' tale captured the imagination of the world. Since then, dozens of similar books and short stories have followed, and feature films with time travel themes remain extremely popular. It's clear that the idea of time travel pushes our buttons in a significant way. It is just too intriguing a concept to ignore.

What is also difficult to disregard are the real life cases of normal everyday people reporting strange encounters with slips, shifts and spontaneous journeys in time. It is also fascinating to note that as modern physics advances, a realization is developing to suggest that time travel may be something that is real and possible, and may already be happening.

The idea of time travel also straddles the fringe theories of New Agers all the way over to hard-nosed science. The world's greatest scientists such as Stephen Hawking and Michio Kaku, are equally enamoured with time travel as is the common man on the street. Support for the possibility that time travel is real runs across a broad spectrum of human thought. Add to this the obvious fact that time travel is a highly entertaining concept - and you get a subject of endless fascination.

We hope readers found this book to be an interesting and worthy contribution to time travel fact and lore. We have attempted to offer stories both from the so-called 'loony fringe' to serious cases that challenge hard science. We've only scratched the surface here, of course, but that is what is terrific about the subject of time travel. It's a source of endless fascination. You don't have to be a gullible believer or a died-in-the-wool sceptic to entertain a subject that has been on the mind of man for countless millennia.

# Book Two:

# More True Time Travel Stories

## Amazing
## Real Life Stories
## In The News

# Contents

# Introduction

Time travel - it's possible. Believe it.

Some of the most intelligent people in the world now agree that it is only a matter of time (no pun intended) before a working time machine will be invented and ready for use.

But there are others who find this laughable, not because they don't believe in time travel, but because they claim to have already done it!

True, some of the latter people are those whom we would consider 'on the fringe'. Yet others have thousands of years of tradition backing them up. Still others are among the top theoretical physicists of our day.

As you will see in the pages of this book, time travel is not merely within the realm of the scientific and technological; it has also long been within the realm of the mystic and ancient religious practices.

What's interesting today is that, for the first time in history, science and mysticism are coming full circle to meet in the middle. Many of the stunning precepts revealed by quantum mechanics, for example, now seem to bear out what the ancient Vedic texts, Buddhists and others have been saying for centuries - that the present way we have come to view time is not the way it is at all.

Since the days of Isaac Newton, who solidified classical physics, time has been viewed as something that moves like an arrow in one direction only - forward. From the present, the arrow of time shoots into the future. The past is where the arrow has already been, and there is never any going back there.

But Albert Einstein's theories of General and Special Relativity changed all that. Einstein himself called time 'an illusion'. Other great thinkers have said that time is not so much an illusion, but works in a much different way that Newton's 'arrow of time' would have us believe. Rather, quantum theory suggests that all time is like a swirling river, with eddies, holding ponds and backward flows and that the present is a 'dynamic moment' with the past and future enfolded into the 'now'.

That may mean that the present is like the 'control point' of time. From the present, it may be possible to go backwards and forwards in time. Maybe we can jump into the river of time and paddle backwards or forwards as we please - if we can just find the key to doing so.

Buddhist and Hindu thought has always maintained the same concept. These traditions view time as an 'eternal moment'. The concept of 'timelessness' is a major element of these ancient traditions, which appear in other cultures as well.

What this suggests is that the ultimate time machine may be the human brain itself. Our place in time - and our ability to move backwards and forwards in time - may be controlled by the fulcrum of the brain, and how it chooses to manipulate the flow of consciousness.

So what is the ultimate method of time travel - the astral travel, meditation and lucid dream practice of the mystically inclined, or the electronic gadgets, wires, electromagnets, lasers and physical manipulators of the scientifically inclined?

In this book, you'll find plenty of examples of both. Then you can decide.

# Lucid Time Travel

For about six months, forty year old Carston Barron of Fort Wayne, Indiana, had been experimenting with his dreams.

About a year earlier he had become fascinated with the idea of lucid dreaming after reading *Exploring the World of Lucid Dreaming* by Stanford University psychophysiologist Stephen LaBerge.

As he began to apply some of the lucid dream triggering techniques LaBerge describes in his book, Carston noticed that he began to dream more frequently and vividly. He found that keeping a nightly dream journal had a powerful consequence, in that writing down as many dreams as possible had the effect of dramatically enriching the content, reality and texture of his dreams. He discovered that it's true to say that when one *pays more attention to dreams, they start paying more attention to you.*

At first Carston's dreams remained of the normal kind. But as time went on, he was eventually able to trigger a lucid dream - and it was such a fantastic experience, he just wanted to go deeper into the practice.

But Carston Barron never expected that his 'hobby' would lead to what he firmly believes is a bona fide case of time travel.

## Background

A lucid dream is a special kind of dream in which the dreamer becomes aware of the fact that he or she is actually dreaming. In a sense, they 'awaken' within the dream. They realize that their physical body is safe and snug in bed, but that their consciousness is now set free to inhabit the dream world.

Once a person becomes aware that he or she is dreaming, the possibilities are endless. In the dream world, the regular laws of physics have been suspended. Anything is possible. You can fly. You can talk to animals; you can even transform yourself into an animal. You can walk the surface of the moon. For that matter, you can travel to any planet, or any exotic location of the universe.

And maybe - just maybe - lucid dreaming is a doorway to time travel.

The question is, could lucid dream time travelling be 'real' time travel, or is it all just a fanciful dream scenario invented by the highly imaginative dreaming mind? Before we delve further into that thorny issue, let's first describe what happened to Carston Barron.

## A Plunge through a 'Time Vortex'

Carston had been practicing a series of specific lucid dream induction techniques for several months with varying levels of success. He had achieved a number of exhilarating lucid dreams in which he enjoyed controlled flight 'like Superman' across beautiful mountainous landscapes. On another occasion he found himself exploring the vast interior of an ancient stone castle.

But on one particular night, he managed to trigger a lucid dream that seemed quite unlike any other. Carston told the American online news site *Examiner.com*:

*'I told myself before falling asleep that I would awaken in my dreams ... it seems that I fell asleep for a short period of time, then suddenly, I felt myself come awake, but I was no longer in my bed! I discovered with total amazement that I was still in my bedroom - except now I was floating four feet above my bed!'*

Rather than a lucid dream, Carston seemed to be having a genuine out-of-body experience. But his adventure was just beginning.

He explained that he only had a little while, maybe less than a few minutes, to try to understand what was happening to him. He found it incredible that he was floating *like a weightless astronaut* in his own bedroom and as he looked below he could see his sleeping body on the bed.

Carston said he looked up towards the ceiling of his bedroom and was stunned to see what he described as a 'kind of swirling vortex composed of pinkish-purple light, with a dark centre'. Even more frightening, he was being pulled into the vortex, and there was nothing he could do to resist.

*'I knew I was going to be taken on a wild ride. I had no idea where I would end up. Normally in the dream state if I get too frightened or excited, I just*

*naturally wake up. But this time it was different. I felt like a leaf captured in a strong river current. I was just being carried along. I was lucid, but I was not entirely in control.'*

Carston plunged into the centre of the vortex. He experienced a 'whining and buzzing sound' as he felt his form being accelerated to an unimaginable velocity. All he could see around him was streaming bits of light intermixed with what looked like black outer space. The sensation lasted for what seemed like several minutes and then - WHUMP! - he 'landed' somewhere. And as he landed he experienced a brief period of disorientation but as he acclimatised to his surroundings he was amazed to see that he was now a boy of about ten years of age - not really himself at all. He also realized that he was now in a classroom with a group of about eight or nine other boys.

Carston said that he and the other boys were dressed in simple smocks, or perhaps one piece tunic-like outfits made of off-white rough cloth, belted at the waist. Some were barefoot, others wore simple sandals. The 'classroom' was made mostly of stone and what looked like dried mud bricks. There were timber beams overhead. They sat on rough-hewn wooden benches.

In the front of the room was a young man who appeared to be perhaps twenty to twenty-five years old who was their instructor. As Carston began to orient himself further, it was clear that he was in a school setting where he was being instructed in some aspect of Greek. The teacher was reciting from memory a

Greek text concerning either religion or philosophy mixed with history.

*'It was incredibly real,'* Carston said. *'It was like I was really there, in a solid physical body. And what was just as strange is that the other boys just accepted me as one of them, like I had always been there. I mean, I felt that I had just crashed into this place from another dimension, or something, but I was just another kid in class. No one seemed to notice me as being anything different.'*

Carston said that even though the teacher and the others were speaking a strange language, he could understand most of what they were saying. He said their speech seemed like a kind of ancient Esperanto - a mish-mash of Greek, Latin and various Middle Eastern dialects, including possibly Aramaic and Hebrew. However, Carston was initially frightened to say or do anything that might draw attention to himself. He was hoping that he would shortly come round from what was happening to him although he was acutely aware that this experience was unlike any other lucid dream he had had before. It felt like he was completely physically in that place - in that classroom. It was so real.

After a short while, the teacher called an end to the lesson and Carston filed out of the classroom with the other children. He went out into a hot, sunny climate. After a walk down a long causeway framed on each side by the high walls of white-stone buildings, they passed into the streets of an ancient, crowded city, before reaching another area - a secluded courtyard with low buildings surrounding a green expanse with fern bushes and small palm trees.

Through one of his classmates, Carlton discovered that he was in the Egyptian city of Alexandria and that the year was 196 AD. In all, he spent just over two days there where he came to know who he now was.

He was an orphan, but a very fortunate one. He learned from the other boys in his class that all of them were orphans, but that they were being sponsored by a wealthy and powerful patron with high political power in Alexandria.

Carston said the other children called him 'Ebius'. He made friends in particular with another boy called 'Bakte'. After some time, Carston became comfortable enough to confide in Bakte that he was having 'memory problems' and that he could not remember his own name, and was having trouble figuring out just where he was and what he was doing.

Bakte seemed to take in his stride the dilemma of his friend 'Ebius'. He filled him in on his situation. He told Carston-Ebius that his father had been a fairly well-known scholar at a local school, and that he had converted to the still developing religion of Christianity. Bakte said that the Emperor in Rome, Septimius Severus, was hostile to Christians and supported a purge of anyone proclaiming themselves aligned to that religion. Thus, the father of Ebius was hauled away and killed in some brutal manner, perhaps by burning.

Ebius was fortunate, however, in that his aunt was a well-connected woman of extreme wealth, and the sister of a certain governor. She took Ebius under her wing and placed him in a learning academy. The other orphans were victims of similar fates, but also had been adopted by benefactors.

Carston said that although he never forgot that his 'true self' was lying asleep in a bed in Indiana in the year 2011, he felt as if he spent two full days, and the best part of a third day, in 2nd Century Alexandria.

Whilst there, he was witness to what he called the 'mind-blowing' culture of the day. He said that on his way to and from his classroom at the academy, he witnessed the average lives of the rank-and-file citizens of Alexandria. He observed levels of existence that were so 'crude, dirty and poor' that he could barely imagine anything that would compare to it today.

Carston said that even though he realized his body was sleeping at home back in Indiana, he also 'slept normally' for two nights in Alexandria in a dorm room for boys, which he said was a 'long sort of shed made of timber'. He said his bed was a 'scratchy wool blanket over straw atop a low wooden pallet'.

Carston said the boys ate communal meals which consisted of a very black, bitter bread and a soup that might have been made with barley or similar grain. He said that although they were young boys, they were given an 'extremely grainy, cloudy beer to drink' that tasted like 'sour dirty water'.

Carston said at the end of the second day he began to feel a sense of panic that he was stuck in this time. Earlier in the day he felt a need to talk about his predicament. He told his friend Bakte and two of the other boys that he was actually a time traveller. He told them that he was a grown man from a time about 2,000 years in the future, and that he lived in a country called America on the other side of the world, across a vast ocean.

The story captivated the other boys, and they asked many questions. Carston said he found it nearly impossible to describe what the year 2011 was like to a group of 2nd Century boys. At the same time, they were enthralled by his descriptions of 'flying chariots' and 'magic wagons' that could carry people along at great speeds. He described other things as well, from the wonders of space travel to great advances in medicine. However, many topics were just too confusing to the boys of ancient Alexandria, such as describing the Internet and robotic space probes.

Carston said he learned much from the boys, as well. For example, they well understood that just gaining the ability to read meant that all of them would become elite members of their society and culture. The ability to read and write was an extreme skill reserved for the very tip of ancient Alexandrian culture.

As for the lessons the boys were receiving, Carston explained that their teacher appeared to be a scholar of great intelligence who taught them the religious and political issues of the very varied cultures of the ancient world. Although Carston tried to listen to what he was being taught, he was beginning to panic that he was going to be forever stuck in that location and was feeling quite scared at the prospect.

His return would happen in the middle of his third night in ancient Alexandria. As 'Ebius', Carston went to his bed. He said he began 'crying to himself' in the dark of the dorm room, and that although he is not a religious man, he began to pray that he would return to his home and 'normal self' back in Indiana in the year 2011.

Shortly after he cried himself to sleep, Carston again came awake and felt the same floating sensation that had originally triggered his journey - again, the swirling violet vortex appeared, and once more he plunged through.

*'The next thing I remember is coming awake in my bed in Indiana,'* he said. *'I can't describe to you what a stunning feeling it was! I jumped out of bed. I was very confused and disoriented for a long time. But I was also exhilarated! The clock said it was 12:30 - I went to bed just after 10:30. I felt like I had been gone for over two days but it had been less than two hours!'*

## Was It Really Time Travel?

In the mind of Carston Barron, there is no question that his experience was far more than a dream, and even something well beyond a lucid dream. He believes he travelled in time, back almost 2000 years to the Roman-ruled era of Alexandria in Egypt. He has no proof, of course, other than *'a personal complete sense of knowing.'*

Sceptics would argue that Carston simply experienced and lived out an extremely vivid lucid dream experience. Many lucid dreamers have been able to set up a dream scenario in which they travel to other times - but they do so in a pre-programmed way.

For example, let's say a person is fascinated with the Middle Ages, the times of knights, lords, ladies and chivalry. Such a person might decide to plan a lucid dream in which he or she would find themselves in such a scenario.

One technique is called 'front-end loading'. This involves a process of inundating yourself with information that will relate to the kind of lucid dream you eventually want to trigger. A person who wants to 'travel' to, say, 12th Century England would spend hours reading books about the period, perhaps watching videos and movies based on those times, including in the hours just before going to sleep.

Then the lucid dream traveller 'sets their intention' to have a dream of 12th Century England as they drift off to sleep. If all goes well, they will dream about that which they have been cramming into their minds for hours or days on end. If a dream is achieved of the 12th Century, the dreamer is tipped off to the fact that he or she is dreaming - they become lucid at that point, and enjoy their dream experience of the 12th Century.

But Carston Barron maintains this is not what happened to him. In fact, he said he had no interest or much of an inkling about the times of ancient Alexandria. He is adamant that ancient Alexandria could not have been further from his mind when he went to bed that night - in fact he maintains that he cannot remember ever reading or seeing any kind of history programme on television about the city. All he knows is that his visit wasn't a dream. As he said:

*'I don't understand the science behind it, or even the New Age philosophy that would explain it. The only thing I know for sure is - I was really there. I'm certain of that.'*

### Astral Travel Connection?

In the case of Carston Barron, the issue of astral travel, or out-of-body travel, clearly becomes part of the issue. Many would argue that at the point Carston found himself floating above his own bed, he was no longer dreaming, but in the out-of-body state.

Those who have practiced OBEs in depth have much to say about the issue of time as it relates to travelling outside the body, and that includes the possibility of time travel. So that's what we are going to talk about next.

# Astral Time Travel

*'You can go anywhere in any time, past, present or future via OBEs.'*
- Robert Monroe, Ultimate Journey

Many people are familiar with the concept of astral travel, today perhaps more popularly called out-of-body travel or the out-of-body experience (OBE). But what you may be less familiar with is how the entire concept of the OBE inevitably leads to the subject of time travel.

What is the connection between an out-of-body experience and time? Why do many feel the OBE is a doorway to the distant past or point in the future? After reading the previous story, you probably already have a clue - but there is more to the story.

Before we delve into the fascinating connection between astral travel and time travel, a little background will be helpful.

## An Ancient and Universal Phenomenon

For centuries, even for thousands of years, people have claimed the ability to travel outside their physical body with some other aspect of their being.

That out-of-body 'vehicle' has gone by many names over the years: the soul, the astral body, the energy body, the etheric body, the higher self, or the soul-self. The general conception has been that human beings have both a physical body which is inhabited, animated or informed and an etheric body or soul.

Different cultures have unique names for what is basically understood as being the same thing. Muslims call it the 'nafs'. From the Vedic tradition we get the term 'atman'. In some Native American languages they call it the 'achak'.

Interestingly, from Chinese religions and philosophy we get two kinds of extra-physical bodies, the 'hun' and the 'po'. The Chinese believed that the 'hun' is the 'higher soul' that ascends to heaven after the death of the physical body, but that another soul, the 'po', is a kind of 'lower soul' that stays encased within the deceased's physical body.

Across all cultures around the globe and over thousands of years, all agree that there is something within, or merged with, the physical body that is non-physical. Furthermore, most cultures and traditions consider it possible for this extra-physical vehicle to leave the physical body behind to travel to locations outside the body, and those locations include a multitude of exotic places. An individual's OBE vehicle can 'pop out' of the physical body to float around right there in the bedroom, just feet from the body - or it can ascend through a series of 'astral planes,' travel to other planets - and also travel to another point in time.

**New Age Renewed Interest**

Although modern mainstream Christianity and other religions have always acknowledged the existence of the soul, the most common belief has been that there is one way and one way only for the soul to leave the body - you have to die. The idea that a person could lie down and 'project' one's soul to other locations was relegated to mysticism, or even forbidden as witchcraft or sorcery.

This, combined with the rise of rational, materialistic science, more or less banished the idea of astral travel to the swampy backwaters of the cultural agenda. Only 'weirdo' flaky New Agers or perhaps the occasional jungle shaman or eastern yogi would dare pay lip service to the idea that one can travel or project one's consciousness outside the body.

But as early as the late 19th Century, there was a resurgence of interest in the idea of astral travel in the Western world, particularly in Victorian era England. The Theosophical Society emerged in 1875 headed by the mysterious Madam Blavatsky. Also in London the Society for Psychical Research was established in 1882 in response to an enormous resurgence of interest in all things paranormal, from séances and ghosts, to ESP and astral travel.

In 1895 a former British Anglican minister, C W Leadbeater, published *The Astral Plane,* a lengthy book which examined in great detail the many facets of what was then believed about the phenomenon of astral travel. A number of other books and publications emerged as well.

Still, such works were scorned upon by the sceptics and scoffed at by the general public.

However, beginning in about the 1960s, the so-called New Age movement emerged in the West, especially in the United States, but also in Europe. With it came an astounding resurgence of all things esoteric. Even so, what was really needed were new authors, new thinkers, and even legitimate scientists who were willing to take another look at what the ancients, across-all-cultures, had been insisting was so for thousands of years.

One of the most remarkable writers to come forward with a series of more than 40 ground-breaking books was Jane Roberts, a minor poet and science fiction writer from a small town in New York. Purely as an experiment, and at the suggestion of her publisher, Jane agreed to write a book on the topic of ESP. Her original goal was to simply do some straightforward experiments and report on her ability, or lack thereof, to demonstrate the idea of extrasensory perception.

Her experiments resulted in an unforeseen development - she began to hear the voice of an extremely intelligent, hyper-intellectual nonphysical entity which identified itself as 'Seth'. Seth claimed to be a 'disembodied energy intelligence' who was obviously extremely eager to speak through Jane Roberts so that he could teach human beings about the true nature of reality.

Jane would simply sit back and recede into a light trance state. Then, Seth would 'take over' her vocal cords and would begin dictating his information, which was written down by Jane's husband, artist Robert Butts. Using this method, Jane began to churn out large books, each hundreds of pages long but, even more remarkable, the information they contained wasn't just more New Age fluff.

Seth was an entity of astounding knowledge whose ideas simply seemed to have an aura of credibility about them for the millions of readers. The collected works came to be known as 'the Seth Material' filling dozens of books and thousands of pages. Jane Roberts, as Seth, never once seemed to contradict herself over decades and millions of published words. Furthermore, large portions of the information made sense when looked at through the lens of modern physics, especially quantum theory. Much of the Seth material contained information about out-of-body travel.

Again, the vast majority of the public and perhaps all of the scientific community, considered the Seth material to be just more of the same - unprovable assertions and New Age trash. The Seth material was dismissed as low-brow books written for cash by a nutty, chain-smoking, petite and frail woman with a knack for turning a phrase. (Jane suffered considerably from arthritis.)

Even so, the impact the Seth material had on the publishing industry was profound. It also attracted the interest of a handful of legitimate scientists. Two of them were Dr Thomas Campbell and his colleague, an electrical engineer by the name of Dennis Mennerich.

Thomas Campbell had recently earned his doctorate degree in theoretical physics. Dennis Mennerich had completed an advanced degree in electrical engineering. As it happens, Thomas and Dennis were both employed in extremely advanced government work. Thomas would later become a high-powered consultant for NASA. But in their early days, he and Dennis moonlighted at an extremely peculiar side job - they worked as volunteers to conduct research for an unusual man by the name of Robert Monroe.

## Monroe Blows the Lid off Astral Travel

Robert Monroe was a wealthy businessman from Virginia. He made his fortune in the radio industry during the 40s and 50s. Robert was in every respect an ordinary, albeit highly successful, entrepreneur. He had absolutely no interest in anything mystical. In fact, Robert was not even a religious man who considered himself an agnostic.

But in the late 1950s, something extremely bizarre happened to Robert Monroe. Without explanation, he began to experience spontaneous and inexplicable vibrations flowing through his body. They felt like currents of electricity. The sensations began one day after he experienced a sudden, extreme attack of pain in his stomach. A visit to the doctors showed there was nothing wrong with him physically. Yet, the stomach pains continued to a lesser degree for many days before they faded - but the strange feelings of electricity flowing through his body kept revisiting him whenever he went to bed, or just laid down for a rest.

He consulted a psychiatrist, who pronounced him 'completely sane' and who suggested he just 'look into' his sensation of electrical vibration to see if he could figure out what his body, or mind, was trying to tell him. One night, while doing just that, something astounding happened - Robert felt his consciousness slip out of his body and pass right through his bedroom floor!

Soon after that, he began to have more out-of-body experiences. A few nights after having what he thought was the 'hallucination of slipping out of his body and falling straight through the floor', he found himself floating near the ceiling of his bedroom and looking down on the sleeping forms of his wife and himself in bed.

Robert suffered through a long period of secrecy about what was happening to him. He even kept the fact that he was expriencing OBEs from his wife. He described it as a time of abject fear and terrible loneliness. He felt isolated by his 'terrible secret'. He simply could not accept his OBEs as something normal or rational. He consulted with more psychiatrists, but they all insisted he was 100% sane. They wrote off his remarkable experiences as 'unusually vivid daydreams' or 'harmless hallucinations'.

The biggest challenge for Robert was conquering himself - his fears, his belief system, his willingness to accept something that certainly must be 'crazy'. Fortunately, Robert Monroe was an intelligent man who had a knack for applying hard logic and step-by-step methods for attacking difficult problems, which had served him so well in the world of business.

He eventually enlisted the aid of the brilliant young but green Dr Thomas Campbell, barely out of graduate school, and others, including an electrical engineer, Dennis Mennerich. They were among a team of six people, which included a psychologist, a teacher and others. At the behest of Robert, they set out to study the out-of-body experience to discover truly what it might be. Was it real? Could others learn to have OBEs? Could a person learn to control the OBE process and induce OBEs at will?

Their eventual findings indicated to them that the out-of-body experience was not only real, but that all people had the natural ability to trigger them if they tried in the right way but, more importantly, if they changed their belief systems to accept that OBEs were not ancient mystical nonsense. They were real. Conquering fear and dumbing old, limiting belief systems was also a major aspect of gaining the ability to 'leave the body' and return safely.

The initial research led to the founding of the Monroe Institute dedicated to the exploration not just of the OBE, but to the fundamental nature of consciousness itself. Out of the early years of research came the invention of something called binaural beat technology, an audio system which people listen to, and which can trigger an out-of-body experience in about 15% of all who try it.

Robert Monroe died in 1995 but left behind a thriving facility devoted to the exploration of all these phenomenon. The Monroe Institute, located in the remote countryside in the township of Faber, Virginia, draws thousands of visitors each year - people from all walks of life who want to experience OBEs for themselves, or explore other facets of consciousness.

In the end, Monroe not only came to grips with his out-of-body experiences, but embraced them. He wrote three best-selling books about his adventures in the 'nonphysical environment'. Among his famous dictums is:

*'I am more than my physical body. Because I am more than my physical body, I can experience that which is greater than my physical self.'*

# More Time Travel

It was clear that after years of research with hundreds of different subjects and using a variety of technologies, that the out-of-body experience was merely the tip of the iceberg in terms of what the human mind was capable of experiencing directly. The range and kind of the information produced by the work of the Monroe Institute and others is well beyond the scope of this book, so we turn now to the subject at hand - time travel.

As Robert Monroe and his team continued to investigate the OBE, a variety of other implications and practical applications came from these studies. The Monroe Institute attracted thousands of people over the next decades, many of whom found they were able to trigger OBEs, and a large percentage of these 'experiencers' reported OBEs that involved what they believed to be actual instances of time travel.

### A Trip One Million Years Back in Time
Robert Monroe wrote about several of his own time travel trips in his last book, Ultimate Journey. In one case, while he was in the out-of-body state, he asked one of his 'guides' in the astral realm to bring him to a time when humanity was no longer living in strife, with greed, war, environmental destructiveness, and so forth - and the guide agreed to lead him to such a place and time.

A moment later, Monroe found himself 'flying' above a beautiful green valley situated between magnificent mountains. He 'landed' and there was conconfronted by a group of human beings who were all naked. They were extremely peaceful and intelligent, and living in complete harmony with Mother Nature.

The people could speak to Robert through direct psychic transference. They were living an idyllic existence in a kind of Garden of Eden paradise. They had conquered all the ills of society, from war and poverty, to greed and hate. Amazed, Robert asked them how many years into the future he had travelled, and how long it would take current civilization to reach this harmonic, blissful existence.

The answer he received was totally unexpected.

The paradise people told him that they were one million years in the past in reference to what Robert considered present day! Robert had not travelled forward in time, but a million years back, to a time when human beings had once lived a halcyon existence. This perfected civilization of two million people said they were ready to 'leave earth for another location'. Robert surmised that after this population left the earth behind, other aspects of the human species would go forward to develop the troubled civilization we have today.

On another occasion, Robert travelled back to a time which he felt might have spanned 'hundreds or even thousands' of years. He found himself in a countryside location where a battle was about to take place. A group of warriors clad in leather tunics belted around the waist were about to confront another group of 'short men with beards'. Both sides were armed with swords and shields. Robert stood by in wonder as a fierce, bloody battle ensued.

Early in the battle, a young man of about eighteen years old was pierced from behind with a spear. The thrust of the spear was so intense, the man fell and was pinned to the ground, the spear sticking straight through his body.

As Robert observed the grizzly scene, he was amazed when he saw the young man struggling to get up - and eventually his 'astral form' popped out of the body. The slain warrior seemed wholly unaware that he was now dead, and he immediately threw himself into the battle again, attempting to slash and kill - only to find that all of his blows passed right through the living!

## The Experience is Repeatable

One might easily write-off the time travel experiences of Robert Monroe as his own vastly inflated and fantastic imagination - except that at the Monroe Institute, the time travel experience has been repeated on numerous occasions by thousands of others who have travelled to the Institute to seek their own experience.

OBEs involving time travel have been replicated by ordinary people, as well as those who have achieved a considerable amount of fame or notoriety in paranormal circles.

## OBE Time Travel Implications

In the vast amount of study that has been conducted on the OBE in the modern era, it has become clear that the phenomenon of astral travel and time travel are inevitably linked. Those who learn to induce the OBE and explore it on a regular basis will almost certainly encounter events of time travel as well.

One of the reasons begins to seem almost obvious - because the way we measure time is inextricably linked to physical matter reality. We can only measure time when it is relative to something physical. Physical objects must be in relation to each other. For example, when we look at a clock, the only way we can measure time is to compare the physical position of the hands of the clock to the numbers on the face. We can only measure the time it takes to travel from Point A to Point B by measuring the physical distance between the points, and you also need the travelling object as a reference point.

But when a person 'goes OBE' they release themselves from physical matter reality. No longer in a 'physical' body, they also become detached from all physical reference points that can be used to measure time!

In effect, in the OBE state, you enter a realm where the ordinary constraints of time flow and measurement no longer applies. The only way to conceive of time, then, is psychologically - with your mind or consciousness.

When you 'pop out' of physical matter reality, you also 'pop out' of the normal stream of time, and this in turn allows you to 're-enter' some location in physical time based on where you want your intent to take you.

This in turn leads to other implications - such as the idea that time as we experience it is largely an illusion. This is what prompted Albert Einstein to say:

*'For us physicists believe the separation between past, present, and future is only an illusion, although a convincing one.'*

Those who practice the OBE almost inevitably come to the conclusion that 'all time is simultaneous' - the past, present and future are all enfolded into an 'eternal now'.

This may be the key to actual time travel, such as we conceive it. Rather than it being the realm of super-advanced technological machines, the ultimate time travel machine - real time travel - may be available to each one of us right now via the process of astral travel - the out-of-body experience.

# Time Machines

As we have seen in the previous stories which explore the idea of time travel via lucid dreaming and astral projection, the idea that one can travel backward or forward through time has been with us for centuries.

But when the human race became scientific, technological and materialist, it was inevitable that attention would be turned to inventing nuts-and-bolts machines that might provide the ability to time travel - and with our 'normal' physical bodies.

The term 'time machine' was coined by British science fiction writer H G Wells when he published his novel *The Time Machine* in 1895. It has remained among the most influential books of the past 100 years. Science fiction writers often inspire scientists to attempt to make real those things that produced wonder and awe in speculative literature.

Today it seems hardly a year or two goes by when new claims of the invention of a time machine 'that really works' come forward, often by an unlikely and obscure source - and these tend to attract a certain amount of media attention. Let's look at some of those now.

### Time Travel in Iran

The most recent example is a man from Iran who goes by the name of Ali Razeghi. In late 2013, Ali claims to have invented a certain kind of time machine. It does not allow for physical time travel, but this device purportedly can read the future up to eight years forward in time. Ali says he can key his machine to individuals and predict everything that will happen to them for the next five to eight years, with 98 percent accuracy.

Ali Razeghi might have been immediately dismissed as just another lunatic except that he is managing director for Iran's Centre for Strategic Inventions in Tehran. He is credited with 179 patents for other working, high-tech devices.

Ali's time-predicting machine can easily fit inside a suitcase, he said. The ability of the device to see into the future is accomplished by a complex series of algorithms which drive calculations that untangle the webs of time. The unit incorporates both hardware and software. Little else is known about the device at this time, however.

Ali's supervisor at the Centre for Strategic Inventions admits that he is 'baffled by the machine' and that this prototype is likely to remain under wraps for at least a year, if not forever. If the device can truly do what its inventor claims it can do, this super accurate future event forecasting device would prove to be of immense strategic value - military, economic and otherwise - to anyone who controls its use.

Further development is likely to be held back, the Iranian scientists say, by the deeply crippled Iranian economy and lack of funds, as well as the ongoing political instability of the region.

Whatever the case, the story of the Iranian time machine went viral and was picked up by just about every major mainstream newspaper in Europe, as well as other media around the world. Sceptics abound, of course, but 'time will tell' if Ali Razeghi's claims are legitimate.

**Small Town Nebraska Man Invents Time Machine**
Another claim for the invention of a time machine that gained considerable and lasting media attention came from a tiny farming community in the American Midwest.

Steve Gibbs was a farmer near the small town of Clearwater, Nebraska. In 1981, Steve said he received a letter which 'just appeared on his table' one day. He explained that he got up and went into another room for just a few minutes, and when he returned, the letter was there.

It turned out to be a detailed note from none other than himself - but a version of himself from the future. It was from this future Steve Gibbs that the 1981 Steve Gibbs received instructions to build a time machine.

Gibbs already had a background in electronics, having attended a trade school. He had gained a local reputation as an excellent 'fix-it' man for all kinds of ordinary electronic devices.

The machine his future self instructed him to build was the Hyper Dimensional Resonator, or HDR. It's a small device, perhaps just a bit larger than one of those old transistor radios. It has three switches, some coils and a magnet. The basic schematics of the device can be found on Steven Gibbs' website.

A person who wishes to time travel simply switches on the device, sets its dials for a specific date, places the coils around his or her head and then rubs a small plate on the surface of the machine. Steve said the maximum time travel effect is achieved only if the person is positioned along with the device over one of the earth's natural 'energy vortexes' or 'ley lines'.

At first Steve claimed the HDR could make you experience only 'out-of-body time travel' but then other people began to come forward who claimed that the device caused them to travel physically forward or backward in time.

Steve Gibbs began to market his device, mostly via buying ads in obscure publications which focused on paranormal topics. He also managed to publish a few articles about his invention, including a story in the *Journal of Natural Health and Parapsychology* in Canada. This led to a few sales, and reports began to trickle in from those who claimed the device actually worked.

For example, a man from Fitchburg, Massachusetts, purchased an HDR from Steve Gibbs. He reported he was able to travel to the year 1945 where he stayed for six hours. In another incident, he travelled to the year 1895 where he spent some time visiting Old Wild West saloons.

Stories like these and others attracted the attention of writer Patricia Ress who decided to investigate the HDR personally. She invited Steve to her home for an interview. He brought along his HDR and Patricia gave the machine a whirl. She did not experience a time travel event, but was unnerved when the device seemed to create 'glowing clouds' and 'strange lights' in her living room. Still, this was far less impressive than being hurled from the present moment into another era of history.

As it transpired, Patricia had also invited a number of other people to her home to meet Steve and fiddle with his time machine. Some were wary of running magnetic energy across their heads via the coils of the HDR, but others were willing to give it a go.

Patricia said the machine was 'running in her home for the better part of a day'. Some of the investigators reported 'psychological time travel' experiences, while others had no results. While slightly unusual, the entire experience with the HDR was underwhelming, Patricia said.

However, it was the next day that Patricia herself came to believe that Steve Gibbs' HDR device was something more than a ridiculous joke or fantasy of a Nebraska farmer. On the following day she sat down to watch one of her favourite movies, the 1953 classic western, *Shane* starring Alan Ladd and Van Heflin. Patricia became quite shaken when she realized that something was very wrong with the movie - much of the dialogue seemed to have been altered.

Patricia had seen the movie so many times she knew the script by heart. But now much of the dialogue had been significantly changed. In many cases,

Alan Ladd was speaking the lines of Van Heflin, and Van Heflin was speaking the lines of Alan Ladd.

She told a radio interviewer:

*'From what I have learned about time is that it can be altered subtly, and in little ways. Sometimes the only way you can tell if there has been a shift or that time has been altered are small changes, or the change of only one thing in your environment.'*

Patricia Ress was impressed enough to devote a full book to her encounter with Steve Gibbs and his HDR time machine. She gave a full account of her experiences in her 2001 book, *Stranger Than Fiction: The True Time Travel Adventures of Steven L Gibbs.*

Since Gibbs first brought out his time machine in 1981, many people have made a purchase from his website and the occasional sensational account continues to trickle into various media reports.

**The Nazi Time Machine**

In the year 2000, Polish journalist Igor Witkowski published a book titled *Prawda O Wunderwaffe (The Truth About the Wonder Machine)*. The information was based on secret files Igor uncovered from sources in Poland. The former classified documents were the transcripts of the interrogation of a top Nazi SS Officer, General Jakob Sporrenberg. He was tried in Poland for war crimes, found guilty and executed in 1952.

Igor's book was later used as the primary source by British aviation journalist Nick Cook for his book, *The Hunt For Zero Point.*

The transcripts revealed that the Germans had been working on a strange, highly technological device which had been dubbed 'die Glocke' which is German for 'the Bell'. The machine was called such because it was bell-shaped - about nine feet wide and twelve to fifteen feet high. It was made from hard, heavy metal.

Inside, the Bell housed two counter-rotating cylinders which were filled with a mercury-like substance that was violet in colour. The Nazis had a code name for a special additive within this substance, which was 'Xerum 525'. In other parts of the machine were employed 'leichtmetall', German for 'light metal', which included thorium and beryllium peroxide.

The device was being constructed in a deep underground facility where V2 rockets were also being constructed. When the cylinders in the Bell were 'spun up' for the first time, there are indications that it immediately caused the deaths of two scientists and other people who were in the vicinity of the machine. It was also noticed that plants that had been in the same area as the Bell were dead and seemed to have 'withered, as with great age'.

The Nazi scientists pressed forward, however, and continued their experiments with the device. The ultimate purpose of the Bell was supposedly an attempt to enable a field of anti-gravity, which would allow German aircraft to become airborne by floating straight up into the air. The Nazis were desperate for a way to launch aircraft because the Allies were having tremendous success in bombing and destroying German runways.

Vertical take-off was the initial impetus behind the Bell technology, but even the Nazi scientists likely did not understand the full implications of what they were dealing with.

As the Nazi scientists forged ahead with the Bell, they discovered it produced other kinds of unexpected phenomenon - one of which was creating a sort of 'window in time'. When a set of mirrors were placed in a strategic location at the top of the Bell, the Nazi scientists were stunned to discover that it opened up a window into events that had happened in the past. In short, it was a kind of television screen that could be tuned to view what was happening in previous years, or centuries.

As sensational as this seems, it is the final fate of the Bell that has generated enormous controversy in recent years - and especially since the revelations of Igor Witkowski's book.

If true, what happened to the Bell is both mind boggling and difficult to believe, and yet the story has gained support in some academic circles, including that of Dr Joseph Farrell, a scholar with the California Graduate School of Theology. The American History Channel also featured an hour long documentary on how the Nazi Bell fell out of the sky - seemingly from outer space - into the American countryside near Kecksburg, Pennsylvania - some twenty years after the end of World War II!

According to Igor Witkowski, Joseph Farrell, Nick Cook and others, the Nazi Bell was used by another high-ranking Nazi SS Officer, General Hans

Kammler, to escape capture by Allied Forces. Kammler was in charge of some of the Nazi's most advanced programmes, including portions of the V2 missile project. Kammler was not a scientist, but a primary administrator for top leading-edge Nazi technologies.

While the history books list General Kammler's fate after the war as 'unknown', some claim that he was captured by the Allies, as were many other German scientists, and was brought to the United States as part of Project Paperclip. It is said that this is how the U.S. obtained many of its best rocket scientists, including the great Wernher Von Braun. Thus, many have theorised that the Bell may have been brought to the U.S. along with General Kammler.

But the more sensational theory is that, rather than be captured by the Allies, General Kammler instead made a desperate attempt to escape capture, and he did so by firing up the Nazi Bell - which was also an anti-gravity device that could fly - but the ultimate result was that he vanished into time. The story gets even stranger when it is said that Kammler came out again over the United States in 1965.

He crash landed the Bell in a wooded area near Kecksburg. The incident prompted what is among the most famous of UFO cases, known as the Kecksburg UFO Incident.

The crashing of the Kecksburg UFO was witnessed by thousands of people. Beginning over the skies of Canada, a bright streak was seen blazing across the heavens. The object passed over Michigan and Ohio, creating a sonic boom which was heard across a three state area. It finally came to rest near the small town of Kecksburg.

Shortly thereafter, a massive military presence moved into the area, including thousands of troops, aircraft and military ground vehicles - clearly the U.S. Military was frantically interested in what had crash-landed in Pennsylvania.

Despite strenuous attempts to keep civilians and the press away from the area of recovery, a number of witnesses saw a large bell-shaped object 'about the size of a small car' extracted from the site. It was hauled away on a flatbed truck under a heavy covering.

There were many specific aspects of the bell-shaped object recovered in Kecksburg that matched the Nazi Bell. For example, witnesses reported seeing a row of what looked like Egyptian hieroglyphics around the bottom of the bell. It was known that the original Nazi Bell bore such markings - partially because the Nazis were known to have put great stock in the symbolism of ancient occult or magical forces - the swastika itself is an example of an ancient 'power symbol' adopted by them.

Dr Joseph Farrell develops his theory in great detail in his book, *The SS Brotherhood of the Bell.* His conclusion is that the infamous Nazi Bell was a device that could bend gravity and therefore alter the fabric of time, and that a desperate Nazi rode the device twenty years into the future, only to crash land to a disastrous end.

Obviously the story of the Nazi Bell is extremely complex, and there are many who vehemently dispute much, if not everything, in the books of Joseph Farrell, Igor Witkowski and Nick Cook. Both sceptics and believers alike agree, however, that a mysterious bell-shaped object was in existence at a top-secret Nazi test site captured by the Allies.

Sceptics say the object was nothing more than an industrial cooling tower, while many others say the preponderance of the evidence shows that Germany's top Nazi scientists - in their desperation to stave off certain defeat - had developed something even more powerful and complicated than they had imagined.

What started out as an attempt to counter gravity may have created a way to bend and create a tunnel through the very fabric of time itself.

# Quantum Jumping

A man by the name of Burt Goldman has been making a considerable career in recent years selling a transformative human development system called Quantum Jumping. Burt Goldman comes from a tradition known as the Human Potential Movement which emerged in the United States in the 1960s although, as you will see, his personal journey is actually much deeper and precedes by several years the Human Potential Movement.

You will also see that the Quantum Jumping methodology, suggested by not only Burt Goldman but others as well, has conjured up considerable implications about the possibility of time travel.

Firstly, some background will help us understand the somewhat complicated connection between quantum jumping and time travel.

## Origins

The central premise of the Human Potential Movement (HPM) is that all human beings possess a vast resource of undeveloped potential. This extraordinary potential can be tapped into using a variety of techniques that combine meditation, visualization, hypnotism, a simple calming of the mind, greater focus of thought and even cultivating mystical states of awareness.

Many agree that HPM has its origins in the ground breaking work of the famous psychologist Abraham Maslow. It was Abraham Maslow who was among the first to break ranks with the dominant schools of psychology of his day - Freudianism and behavioural psychology.

Maslow is sometimes called the 'Anti-Freud' because he suggested that Freud's theories of what drives human behaviour were backwards. That is, he said that Freud studied the mentally unstable and the mentally ill, and then applied the dysfunctional minds of his patients to the entire human race.

In opposition to Freud, Abraham Maslow studied mentally healthy, happy, successful and highly accomplished people. He looked for what was the best in people - and then suggested that every human being has the same potential to evolve towards higher states, to grow and be happy. Rather than taking the Freudian approach of assigning this 'complex' or that 'dysfunction' to all humanity across the board, Maslow looked first for what was best in people, and suggested each individual build on that.

Likewise, Abraham Maslow was critical of the great B F Skinner, the founder of behavioural psychology. This did little more than reduce human beings to the level of robotic animals - every behaviour was rooted in simple action-reaction to stimuli encountered in the environment. Higher thought was viewed by behaviourists as a kind of delusion created by the normal physical functioning of people to obtain the basic needs of life, from food to sex, to shelter and defence.

Maslow observed that many aspects of human consciousness could simply not be pegged to the action-reaction process of physical actions or behaviour. He observed that people who were in a bad place in their lives could be spring-boarded to a higher state in their lives by uncovering their true potential. For whatever reason - a bad childhood, poverty, a negative attitude, or just bad luck - the many powerful abilities latent in every human being had become repressed or buried. The key was to find ways for all people to get rid of what was blocking them or holding them back, so they could move forward and grow.

**The Silva Method**

By the 1960s the ideas of Abraham Maslow had expanded and been taken up by a variety of other people and disciplines. There was a new excitement in shaking off the dreary theories of Freud, and cutting the chains of behaviourism. Instead, the focus was on the idea that every person is already in possession of unlimited potential. It was all within them, in their human consciousness. What was needed were methods to find, uncover and unlock all that hidden potential.

One of the most influential tools of the Human Potential Movement came from an unlikely source - a remarkable man by the name of José Silva. José was an orphaned Mexican boy who grew up in the border regions of Texas and Mexico. He had almost no education, other than a few years of basic grade school. He was definitely born onto one of the lowest rungs on the ladder of American society.

José, however, from an early age displayed the skills of a natural entrepreneur. As a boy he sold newspapers, shined shoes and did odd jobs for cash. He quickly moved into selling other merchandise door to door - and hired other youngsters to work under him for commission. Before long he had a small army of sales boys. By the time he was in his early teens, he was making more money in a day than most adults earned in a week or a month.

In the 1940s Silva began to study the human brain and human behaviour. He also began to develop a method whereby people could vastly increase the power of their minds by making better use of 'both halves of the brain'. The method he eventually developed involved calming the mind with a simple meditation process, creative visualization and more. He eventually formalized a specific set of steps and techniques which came to be known as the 'Silva Method'.

At first José Silva intended only to use his human development methods for his own family, mostly for the betterment of his own children. But word began to spread about the mysterious new Silva technique. He began to conduct seminars and teach his methods to others and this, in turn, led to even wider interest in the Silva Method.

Even so, José Silva did not begin selling a formalized version of his Silva Method as a package until well into the 1960s. But when he did, the programme rapidly achieved worldwide recognition and success.

As much success as Silva enjoyed with his programme, something else would happen in 1980 that launched the Silva Method to an even higher level of achievement - it was in the late 1970s that José Silva met a man by the name of Burt Goldman.

## The American Monk

Burt Goldman was nineteen years old and deployed with the U.S. Military in Korea during the war from 1950 to 1953. Whilst there, Burt Goldman met a man by the name of Kwan Jung, a kind of oriental shaman who was adept in the ancient mystical practices of working with Qi energy. This is the 'life force' or 'natural energy' that is at the basis of all that is living, including the human body.

Burt learned to manipulate his own Qi energy under Kwan Jung, and from there, continued to probe deeper into the mysteries of ancient energies that seemed to be at the core of all human life. After the war he went to Hawaii where he explored a similar aspect of Qi energy under a native Hawaiian 'Kahuna Master'.

Back in the mainland United States, Burt Goldman became a student of the great Indian Yogi, Paramahansa Yogananda, author of the influential book, *Autobiography of a Yogi*. It was from this teacher that Burt probed deeper into the art of consciousness exploration by learning formal meditation based on the many ancient traditions of India.

Burt Goldman eventually began to teach his own classes and seminars and to publish articles and books about what he had learned about human potential. Before long, he was dubbed 'The American Monk' by the many people he taught through the 1950s and 1960s.

By the time Burt Goldman met José Silva in 1979, both men were well-positioned to come together for a remarkable synergy of purpose and energy. Burt Goldman found that everything he had learned over his past decades of consciousness work could be expertly packaged within the structure of the Silva Method.

It took Burt only months to become the No. 1 Silva Method instructor in the United States. In 1989, José Silva and Burt Goldman co-wrote a book titled, *The Silva Mind Control Method of Mental Dynamics*, which became an instant best seller.

### Quantum Jumping

Burt Goldman's career, beginning when he was a green 19-year-old sergeant in the U.S. Army, reveals a man who never rested on any particular method or tradition. After learning all he could about Qi, he explored the 'huna' of the Hawaiian shamans, then Vedic meditation. He went on to explore various forms of hypnotism and remote viewing, and gave full application to the methods of the Silva approach.

A few years ago, Burt became fascinated with the astounding implications coming out of modern quantum theory. One particular aspect of quantum theory was proposed by the theoretical physicist Hugh Everett III in 1957 - the Many Worlds Interpretation.

## Infinite Parallel Universes

In short, the Many Worlds Interpretation suggests that there is not one universe or one dimension - but many. In fact, there are literally an infinite amount of other universes existing parallel to our own. Hugh Everett developed the Many Worlds Interpretation to solve vexing problems that were cropping up in the overall model of quantum theory. He said that the idea that there are millions of other universes was the only way to explain the many bizarre phenomenon suggested by quantum mechanics.

Each NEW universe is created every single time a human being makes a choice. A very simple example:

You get up in the morning and have a choice between coffee and tea as your morning wake-up beverage. Let's say you choose coffee. If you do, then an entire universe is created to accommodate your choice to drink coffee. But because there was another choice - tea - there will be another universe where you choose tea instead of coffee. As difficult as it is to believe, an entire universe will be created for the 'you' who selected tea over coffee.

Now imagine all the dozens, if not hundreds, of choices you make every day, large and small. Every time you choose one thing over another, an entire universe is created to accommodate each choice.

What this means is that whole universes are in a constant state of splitting away from each other, each branching off to go its own path. It also means that there is a 'duplicate you' in each of these other universes. There are millions and billions and trillions (in fact, an infinite number) of parallel universes, and a version of 'you' inhabits each one.

Furthermore, as each individual universe branches off and goes its own way, they become increasingly different from each other. These differences are caused by all the various choices each 'duplicate you' continues to make as time moves forward. This can lead to major differences in how each of your duplicates live out their lives.

For example, let's say that back when you were deciding on a major for college, you were torn between computer science and an artistic field, such as photography. Well, in the Many Worlds Interpretation, one of you would choose computers and the other would set off on a career in photography. Still other 'yous' might choose different paths - one will go into medicine, another into journalism, another accounting - and so forth.

**Tap into your Doppelganger**

What Burt Goldman is suggesting is that Hugh Everett's Many Worlds Interpretation is in fact a reality - and a recent survey of top theoretical physicists today revealed that more than 80 percent of them also accept Many Worlds as fact. This includes such luminaries as Stephen Hawking, Michio Kaku, Fred Alan Wolf and many others.

Burt Goldman calls each of the infinite number of 'other yous' in all of those other universes your 'Doppelganger', borrowing the term from German lore. In German mysticism, a doppelganger was a kind of paranormal spirit double of a person which could appear at odd times for reasons not well understood.

Burt's idea is that by using various meditations, visualization and self-hypnosis techniques, any one of us can actually get in touch with any one of our other doppelgangers in their own universes. Furthermore, Burt says that by contacting or communing with your alternate-world doppelganger, you can tap into all the skills and talents this other you developed as they went their own way in their own world.

For example, let's say you are a carpenter who never studied music, but you decide you want to learn to play the piano. Using Burt Goldman's Quantum Jumping methods, he says you can actually find another you - a doppelganger - who actually chose to devote his or her life to music and playing the piano. Because that person is actually another version of yourself, it is possible to tap into the life skills the other self has developed and incorporate them into your own life.

Burt says that tapping into the unlimited number of doppelgangers you share a common origin with, can enable you to easily absorb or obtain just about any skill, knowledge or capability your heart desires.

## Time Travel Implications

Okay, but how does time travel come into this already astounding picture? Well, the fact is the very nature of the mechanics of the Many Worlds Interpretation is deeply tied to the concept of time, and how time works in the greater scheme of things.

A recent mathematical discovery by David Deutsch of Oxford University has been said to confirm once and for all that Many Worlds is correct and parallel universes are a reality. Furthermore, David said that the implication for time travel - and the potential ability to successfully time travel - are profound.

David Deutsch told a reporter for the British newspaper, *The Telegraph*, in 2007, that parallel universes provide a kind of physics loophole that makes time travel possible:

*'It does sidestep it. You go into another universe,'* Deutsch said, though he admits that there is still a way to go to find schemes to manipulate space and time in a way that makes time hops possible.

*'Many sci-fi authors suggested time travel paradoxes would be solved by parallel universes but in my work, that conclusion is deduced from quantum theory itself,'* Dr Deutsch said, referring to his work on Many Worlds.

The article goes on to say:

*Dr Deutsch showed mathematically that the bush-like branching structure created by the universe splitting into parallel versions of itself can explain the probabilistic nature of quantum outcomes. This work was attacked but it has now had rigorous confirmation by David Wallace and Simon Saunders, also at Oxford.*

## Slip-Sliding Across Universes and Time

Now, if grasping the idea of multiple and parallel universes is not mind boggling enough, one must still consider this added dimension to the picture - the element of time travel.

To date, thousands of people have worked with Quantum Jumping and they claim to have had success in meeting their alternate universe doppelgangers. However, many report a strange additional twist - they appear to have transported themselves to another era of time. That's because when people find their duplicate selves in other universes, they often discover they have contacted a past self or a future self.

For example, many quantum jumpers say they have 'jumped back' to a time when they were a teenager, and have been able to prevent themselves from making a serious mistake - altering the timeline of their present and 'home base' universe for the better.

Other quantum jumpers report even more dramatic time travel journeys. Some find themselves talking to a duplicate self who is living in medieval times, while others confront a version of themselves who appear to be living in a far-future world.

Others not associated with the Burt Goldman technique of Quantum Jumping also report frequent and numerous time travel experiences. One such person is the popular author Cynthia Sue Larson. In her book,

*Quantum Jumps,* Cynthia describes dozens of time travel experiences when using quantum jumping practices. She has been featured on a number of prime TV programmes, including the History Channel's *Weird or What.*

Like Burt Goldman, Cynthia Sue Larson recognized the Many Worlds Interpretation as a reality and something that does not necessarily belong only to the realm of the complex mathematics of physics, or as a point of pure belief. She says anyone can prove to themselves - through direct personal experience - that quantum jumping and quantum jumping time travel is not only possible, but a reality.

Cynthia herself has a degree in physics and explains that 'all time is symmetrical'. She said that entering a parallel universe at any point means that you are entering it at a specific point on its timeline - and that you can also slide up and down that timeline.

**Reality is Strange, but it's still Reality**
The famous British biologist J B S Haldane said:
*'The Universe is not only queerer than we suppose, but queerer than we can suppose.'*

For many people, something as incredibly bizarre as multiple worlds, parallel universes and the fact that duplicates of ourselves are inhabiting those universes, is a lot to swallow.

It's not surprising then that Burt Goldman has, in the course of promoting his concept of Quantum Jumping, garnered armies of critics and sceptics denouncing him as the worst kind of fraud and charlatan.

Goldman is also often vehemently denounced for the aggressive marketing efforts he uses to promote and sell his Quantum Jumping training course. For many, Goldman's claims seem outlandish and smack of pseudo-science. Many denounce him for his use of the term 'quantum', which they claim is an improper or misleading use of a scientific term.

On the other hand, for thousands of people, the proof of the pudding is in the eating. They've tried Quantum Jumping, they say it works, and they attest to the positive changes it has made in their lives.

They also report frequent experiences of time travel. Ultimately, perhaps the only way for each of us to decide is to not simply believe what both the believers and sceptics have to say - but just try it for ourselves.

There's no replacement for direct experience. And who knows, you may even find yourself taking a journey through time.

# Time Travel Tomb

True tales of time travel probably cannot get any stranger than that which is associated with the supposedly brilliant but highly eccentric British inventor, Samuel Warner.

Dead for more than 150 years, many have come to believe that a uniquely constructed tomb in London's Brompton Cemetery may actually be a time machine of Warner's design. He built it with the help of a respected Egyptologist and architect of the day, Joseph Bonomi.

The story of Samuel Warner has all the elements of a fantastic Hollywood movie, or perhaps one of Stephen King's best horror-fantasy novels. Warner's story involves Egyptian tombs, a royal mistress, and a 'maverick genius' scientist who may have been murdered to prevent his various secrets from getting into the wrong hands. For good measure, there is a lost key to a strange building with no floor plan and which, on paper, does not and should not exist.

The story even involves a bizarre plan to set up a series of seven teleportation devices throughout London cemeteries in the 1850s. It sounds impossibly strange, yet it's true.

### Who was Samuel Warner?

Samuel Alfred Warner was born around 1794. His father was a sea captain and owned a small ship called the Nautilus. Warner was known to be an inventor, according to the British Naval historian, John Knox Laughton.

John Laughton wrote that Samuel Warner began 'pressing' the admiralty to purchase the design for a device he called 'an invisible torpedo' in 1830. The device was small, the size of a duck egg. Samuel claims that he had demonstrated the power of his invisible torpedo to great effect by sinking two ships while serving on his father's ship, the Nautilus, during the war of 1812.

This was all well and good, except Samuel refused to hand over the design plans to a committee of British naval ordnance experts who were charged with investigating weapons of potential interest.

Samuel would not show anyone the tiny invisible torpedo itself - although a spectacular demonstration was presented in which a ship was apparently blown to smithereens, to the astonishment of naval observers. But how did the ship really explode? Was some manner of clever fraud involved?

Samuel Warner refused to show the actual device to anyone in the navy, nor would he release design schematic unless the navy first forked out £200,000, which would be the equivalent of about £7 million today.

Samuel also had an aerial version of a ship-sinking bomb, designed to be dropped from a balloon. He wanted £200,000 just for a peek at that too, so he was asking for about £14 million from the admiralty based solely on his claims, and perhaps at least one spectacular demonstration.

There is reason to believe Samuel Warner was taken somewhat seriously by the British admiralty. In 1842 records show that a high-powered panel headed by Sir Thomas Byam Martin and Sir Howard Douglas determined that, during the War of 1812, the ship owned by Samuel's father served an important military function. Its mission was the transportation of spies. There was also evidence that this small ship somehow destroyed at least two enemy vessels.

Records show that in 1842 Samuel's duck-egg-sized torpedo was still being investigated after more than 10 years. Finally, the Duke of Wellington was involved in investigating the weapon in cooperation with the navy's ordnance department. However, it all came to an abrupt end in 1853 when Samuel Warner 'died under mysterious circumstances'.

Samuel was supposedly buried in Brompton Cemetery, although some records show no corpse was ever recovered. Samuel's grave is unmarked but, even so, the plot may lie empty. He left behind a wife and seven children.

The matter might have rested there, except subsequent investigation by high-ranked military authorities and local police uncovered a number of strange incidents surrounding the activity of Samuel Warner.

## London's Mysterious Egyptologists

As it happens, Samuel had been a close friend to one of London's most respected architects and Egyptologists, Joseph Bonomi the Younger. His father was Joseph Bonomi the Elder.

Originally from Italy, Bonomi the Elder moved to London as a young man. He spent the rest of his life in England and became a powerful and respected architect. He was also known to have been fascinated (some say to the point of obsession) with the culture of ancient Egypt.

One of the most famous architectural designs of Bonomi the Elder was built in 1796 and stands today on a country estate near Norfolk in the village of Blickling. The structure is a mausoleum in the shape of an Egyptian pyramid. In it are interred Sir John Hobart, the 2nd Earl of Buckinghamshire, and his two wives.

Today the Blickling pyramid is a favourite site of investigation by paranormal investigation clubs. Reports of strange phenomenon abound in proximity to the odd Blickling monument. Everything from UFOs and aliens to ghosts and strange creatures have been spotted in the vicinity where the pyramid stands in wooded seclusion.

It's clear that Bonomi the Elder passed on his great interest in ancient Egyptian culture to his son, Joseph Bonomi the Younger. It prompted the younger Bonomi to spend nine years in Egypt conducting first-hand archaeological studies and painstaking examination of Egyptian hieroglyphs and other arcane symbolisms. A curator at the British Museum said later that Bonomi was the most knowledgeable Egyptologist in all of Great Britain.

It is known that Joseph Bonomi the Younger was among the first to lay his eyes upon certain papyri scrolls containing voluminous hieroglyphic texts found in the Valley of the Kings. Many believe to this day that Bonomi learned incredible information from these papyri scrolls that were kept a deeply guarded secret - including possibly the key to a method for teleportation and time travel.

## The Inventor and the Egyptologist

What was the nature of the relationship between Samuel Warner and Joseph Bonomi the Younger? It is difficult to say, but what is certain is that they knew each other well and came together in an area of mutual interest.

Bonomi was of the upper class, based on the renown of his father, but also because of his own significant achievements in painting, Egyptology, architecture and more. He was married to Jesse Martin, the daughter of one of England's greatest and most wealthy painters, John Martin.

Samuel Warner on the other hand was a commoner who struggled. It is known that he died in poverty. He left behind a wife and seven children whose fate was grim after the mysterious death - or as some say, 'the vanishing' - of their father.

Whatever the case, the paths of Bonomi the Younger and Samuel Warner came together and the two men formulated a plan to build and distribute a series of teleportation 'booths' in strategic locations around London.

## A Bizarre Plan

The 'inventive genius' of Samuel Warner and the secret Egyptian knowledge of Joseph Bonomi would come together in the hopes of streamlining transportation in England's biggest city, which was famous for its congestion and unbearable traffic. Getting from to any Point A to any Point B was a certified nightmare.

To add an element of strangeness to the teleportation system plan, the two men resolved to build each 'booth' in a graveyard. Their reasoning was pure practicality.

Beginning in 1839 a series of new graveyards were established in strategic locations around London. Once a graveyard was established, it was a permanent facility - unlike other sections of city blocks of the city which could be subject to constant change, reconfiguring, rezoning or re-routing of streets, and so forth.

A cemetery was also a convenient place to build unusual structures. A graveyard was a good place to build something out of the ordinary while being undisturbed by observers. In a graveyard, highly eccentric structures could be explained away as the strange last wishes of the dead. A graveyard is a location where a special structure was unlikely to be disturbed over long periods of time.

Thus, the teleportation booths designed by the 'maverick inventor' leveraging the ancient Egyptian knowledge obtained from the papyri scrolls in the Valley of the Kings by Bonomi the Younger began working their magic on the London transportation system from their graveyard stations.

## The Strange Connection of the 'Three Spinsters'

You may be wondering by now what these teleportation devices looked like, if they were really placed in graveyards throughout London and on just what principle did they work? You might also be wondering if these devices could serve as time travel machines.

All in good time. But first we must make mention of a major part of this already strange story.

Even though Joseph Bonomi the Younger was somewhat well-to-do, it is also known that his income was 'somewhat uneven'. He was either unwilling, or not able, to provide the financing for such an ambitious project as placing seven Egyptian teleportation devices among seven London graveyards.

To this end, Joseph Bonomi and Samuel Warner used their considerable charisma to charm an aging, wealthy woman into bankrolling the project. Her name was Hannah Courtoy. She was a woman whose life conjures up considerable mystery and drama.

In brief:

Records show that Hannah Courtoy was born Hannah Peters between 1780 and 1782. It is believed that she was the mistress of a wealthy man named John Courtoy. It is widely acknowledged that John Courtoy fathered three children with Hannah: Mary, Elizabeth and Susannah.

John Courtoy's original name was Nicolas Jacquinet. He was born in France and was a wig maker by profession. Wigs were a big deal in the 17th Century, so in this line he managed to accumulate a large fortune.

At some point, Nicolas Jacquinet came to be known as John Courtoy when he took up residence in London. It is also known that he was at least fifty years older than his mistress, Hannah Peters, with whom he fathered three children.

John Courtoy died unexpectedly in 1815. Although a man of considerable wealth, he left no will, dying intestate. It is almost certainly true that John Courtoy was not legally married to Hannah Peters - although upon his death, Hannah soon began calling herself Hannah Courtoy.

A French woman - who claimed that her mother was John Courtoy's half-sister - agreed that Hannah, despite never being married to John Courtoy, had best claim to his fortune as she was the mother of three of his children.

It is known that Hannah (Peters) Courtoy hired a certain 'shady agent' to whom she paid £300 to concoct a phony will. The will was adjudged to be a fraud in court, yet Hannah prevailed in receiving a sizeable cash settlement from the estate of John Courtoy. She also received diamonds, plate, a collection of fine china and linen, but did not succeed in her attempt to retain possession of John Courtoy's house.

Nevertheless, Hannah managed to use her wealth to purchase a home at Wilton Crescent, Belgravia

Square, London. There she lived with her three daughters, none of whom ever married, and thus came to be known as 'The Three Spinsters'.

Between 1848 and 1852 a plague of cholera swept through the cramped city of London, claiming thousands of lives, including that of Hannah Courtoy. Official records list her death on 26 January 1849.

Before her death, Hannah Courtoy had obviously formed an association with Samuel Warner and Joseph Bonomi the Younger. The two men were successful in convincing her to provide the funding for their plan to build seven ancient Egyptian technology teleportation devices in key graveyard locations throughout London.

Some writers and researchers speculate that in exchange for her financial patronage, Samuel and Joseph promised to build a special kind of tomb for Hannah Courtoy and her three daughters - a tomb that would allow them to cheat death by transporting them to another location in time.

Unfortunately, Hannah died four years before the time-portal tomb designated for her was built and erected in Brompton Cemetery, London. Her body was supposedly moved from its original place of burial to the tomb in 1852. It is believed that two of her three daughters were also interred in the special tomb.

## Are the Courtoy Women Really in the Tomb?

There is reason to believe that Hannah Courtoy and her two daughters are not resting in their elaborate Egyptian tomb in Brompton Cemetery.

For one thing, the tomb was last opened some 120 years ago, locked, and a key for the door to the tomb has never been found. The key disappeared under mysterious circumstances.

Secondly, some believe the Courtoy tomb of the 'spinsters' is not a tomb at all, but one of the teleportation devices that Samuel Warner and Joseph Bonomi the Younger set up in London Cemeteries.

Adding to the evidence for the teleportation device theory is the strange fact that no registry for the Courtoy tomb from Brompton Cemetery exists - an enormously strange circumstance since all gravesites and tombs in London cemeteries are well-known to be painstakingly recorded in public graveyard records.

London officials are famous for their extremely fastidious documentation of gravesite locations, and especially for that of elaborate monumental tombs. In fact, no permission could be granted to erect such a monument in any cemetery unless a large body of bureaucratic red tape had been completed first.

No records have ever been found for the tomb which supposedly houses the bodies of the Courtoy women. On paper, the tomb does not exist. Without paperwork, some historians say it would have been simply impossible to place a tomb in that location.

Many today believe the Courtoy 'tomb' sits empty, and that the three Courtoy women are not buried there - and in fact it is not a tomb at all, but rather a gateway to travel elsewhere in time.

The Courtoy women were immensely wealthy at the time of their deaths - yet they left behind no will, and no public records of their existence.  The final fate of their fortune has never been recorded in any probate court in London.  It seems that the Courtoy women either died or vanished, as did their vast fortune.  Did they take it with them to another location in time?

## The Time Tomb

The Courtoy tomb is a vertical trapezoid constructed of dark polished granite.  It is fitted with a large door made of bronze.  Carved around its exterior are a series of Egyptian hieroglyphics supplied by the knowledge and expertise of Joseph Bonomi.

At least three of the 'time tombs' still survive - the one in Brompton, a similar structure in Highgate and another in Kensal Rise.  Like so many other things that seem to vanish in this strange tale, the other four 'time tombs' or teleportation devices or whatever they were, have also vanished from their original locations - despite having been massive structures built of heavy stone and bronze.

## Did Samuel Warner Escape into Time?

As we have said, the death of 'maverick genius' inventor Samuel Warner was unusual, if not suspicious.  His death is recorded as having occurred in 1851, which would have been in the midst of the chaos created by the horror of London's cholera plague.  It would not be unusual for a man living in or on the edge of poverty to die of cholera and be committed to a hastily prepared, unmarked grave.

Conspiracy theories abound, however. For one, some speculate that Samuel's torpedo and aerial bomb inventions are what led to his death. As we said, Samuel was demanding a total of £400,000 from the British government for these devices, and he was being taken seriously.

When negotiations broke down for the rights to his deadly torpedo, he may have threatened to offer it to a foreign government instead, which is something the British government did not want to happen. The suggestion he was assassinated by government officials is highly credible in the eyes of many.

Another popular theory among Victorian England, however, was that when Samuel Warner's troubles became too great, he opted out and used one of the teleportation devices to escape into another time.

### The Evidence of the Grave

Joseph Bonomi the Younger not only escaped the London cholera plague, but went on to live to the age of eighty-two, dying in 1878. Many point to clues supposedly left behind in his tombstone.

Joseph is buried just a few feet from where the 'time tomb' or the mausoleum of The Three Spinsters stands. His tombstone is rather simple for a man who was wealthy and distinguished.

Twenty yards away from the time tomb, Joseph's gravestone bears similar hieroglyphic carvings to the tomb. It is also adorned with an image of the Egyptian god of the dead, Anubis, who is depicted sitting on what appears to be a replica of the mausoleum.

Many believe this is a vital clue to the time tomb's secret. The direction Anubis is facing - towards the mausoleum - suggests in Egyptian mythology that a soul has been 'lost in time'.

Interestingly, Joseph's grave marker was not set up upon his death - but at the same time that the time tomb mausoleum was completed. Joseph would not be buried there until more than a quarter of a century later.

**The Tardis**

It is often said that the exotic tomb of the Three Spinsters in Brompton cemetery, together with those similar structures that exist at the other sites, are the inspiration for the time-travelling TARDIS of the popular television show, Dr Who - but even this is controversial.

TARDIS stands for Time and Relative Dimension in Space. In the TV series, the TARDIS resembles a police telephone box. Unlike an ordinary callbox, the police box telephone is located behind a hinged door so it can be used from the outside, and the interior of the box is, in effect, a miniature police station for use by police officers.

The TARDIS is similar in size, shape and function to the monument in Brompton Cemetery.

Some say it is pure balderdash to suggest the TARDIS was inspired by the graveyard teleportation devices that were supposedly designed and built by Warner and Bonomi. The first Dr Who episode was aired in 1963 and the creators make no mention of a tomb designed by Joseph Bonomi.

Yet others maintain that it is well known that the people of Victorian England widely believed that the ancient Egyptians had mastered the ability of time travel. The creators of Dr Who could have easily run across the story of Samuel Warner, an eccentric genius who may have been the model for Dr Who himself. Add to this the size and shape of the TARDIS, and it seems possible that inspiration for the Dr Who series came out of the strange tale of Warner, Bonomi and their teleportation devices.

## A Timeless Mystery

So the story of Samuel Warner and the possibility he collaborated with a renowned Egyptologist to construct a series of teleportation time machines strategically placed about London is one of endless twist and turns.

Did Samuel Warner truly vanish into time? Was he murdered to keep his secrets from getting into the wrong hands? What happened to the key for the Egyptian time tomb mausoleum? Why do no records or plans for the structure exist? What happened to the other 'teleportation' devices? How was Hannah Courtoy persuaded to part with a great deal of her wealth to support the project? Do the women rest in the tomb - or did they 'teleport' to another century and location in time? What happened to their fortune? Why do no records of their lives exist? Was Joseph Bonomi attempting to leave behind clues with the symbols on his tombstone?  Did he really possess knowledge of Egyptian time travel obtained from pap-ryi found in the Valley of Kings?

We may never know the answers, unless of course the lost key to that strange tomb in Brompton Ceme-tery is found one day, and someone dares to open it.

# Mallett's Time Machine

One of America's top physicists has a message for humanity:

*'I want the public to know that time travel is possible ... and I think a working time machine can be built in about ten years.'*

This might be just another wild claim from some crackpot Internet ranter, but in this case, the statement is made by Dr Ronald L Mallett, a highly respected professor of theoretical physics at the University of Connecticut.

Ronald Mallett has been driven to solve the riddle of time ever since he lost his father to a heart attack in 1955. He was just ten years old when Boyd Mallett died at the too early age of thirty-three.

*'He was the centre of my life,'* Mallett said. *'For me, the sun rose and set on my father.'*

The senior Mr Mallett was a heavy smoker, frequent drinker and kept an otherwise unhealthy lifestyle. When young Ron Mallett learned that his father might have lived a longer life by dropping his bad habits, he became obsessed with the idea of time travel. He wondered if he could find a way to go back in time to warn him - to help him change his ways and be there for his four children.

**A Journey Out of Poverty and Prejudice**

Young Ronald Mallett had a lot going against him.

He was black in a time of rampant racism in America. His family made great sacrifices to move away from the dangerously hostile-to-blacks environment of the Deep South of the 1950s to re-establish themselves in New York City. There they found a new start, but were still outsiders. The Malletts lived in a working-class, mostly Jewish neighbourhood.

But his father had a special talent. Boyd Mallett was a certified gadget freak. He loved all things electronic. He was also smart. His passion for electronics and willingness to work hard found him good work - including wiring the new United Nations building, then under construction in New York.

Boyd Mallett spent hours of his spare time interacting with his young son, who shared his father's love and fascination for electronic gizmos. Father and son would spend endless hours huddled together in their garage cobbling together all manner of contraptions with capacitors and circuits, wires and crystal set components, and more.

When Boyd Mallett died suddenly, it shattered Ron's world. The ten year old boy sank into a deep state of despair that became a lasting depression. Part of his escape was browsing science fiction pulp magazines and comics but he soon moved on to reading full-length science fiction novels.

It was one particular work of science fiction that captivated Ron Mallett - *The Time Machine* by H G Wells. Ron Mallett became 'electrified' with the idea that someone might be able to travel into the past because, he reasoned, if you could transport yourself into an earlier time, you would also see again the ones you loved and lost.

You could also warn them!

*'What if,'* Ron Mallett thought, *'I could build a time machine, go back and meet my father, and tell him to stop smoking and drinking because it would soon kill him?'*

For most people, this would be little more than an unproductive fantasy, but Ron Mallett had something special going for him. He had inherited his father's genius for electronics. That, combined with a passion to achieve something special, drove him forward.

### Another Significant Death

Ron Mallett's father died in 1955, and the young science fiction and electronic fan could not help but take special notice of the death of another great man who also died in 1955 - Albert Einstein.

Einstein had spent his life blowing the lid off the world of physics by showing that time and space were not actually separate qualities of the universe, but actually one and the same thing. Time and space were inextricably connected. Einstein gave us the concept of 'space-time' which became the core of how physicists and cosmologists have developed our most fundamental models of reality ever since.

Ron Mallett admitted he became obsessed with Albert Einstein and his ideas. That was because he realized that this new way of thinking about gravity, space and time opened a doorway for him. That doorway led to the possibility that someone could build an actual bona fide time machine. The new physics showed that it was definitely possible.

Einstein's equations revealed that if you could figure out some way to twist the fabric of space-time around, it may be possible to form a loop - one end would connect the past to the other end, the future. Next, a way was needed to 'enter' that loop, or jump along for the ride into another time.

## Obstacles

Ron Mallett had a lot going against him. He was fatherless, poor and black. He was one of four children that his widowed mother struggled to raise on the minimal income of a window cleaner. The idea that she could ever afford to send any of her children to college was, quite frankly, an impossibility.

Thus, after graduating high school, Ron joined the U.S. Air Force. He hoped his military service would help him earn a scholarship to college. Once in the Air Force, it became quickly apparent to Mallett's superiors that here was a young man who was not only very smart, but he also possessed an incredible knowledge of electronics.

He was fast-tracked into the Air Force electronics training programme.

Unfortunately, Ron's first tour of duty was in Biloxi, Mississippi, one of the most deeply racist areas of the American South. He had returned to a location that he and his family had struggled so hard to escape several decades ago.

Black people were at such risk in the Biloxi area that Ron Mallett was advised to never leave his military base. If he did, a beating or even a violent death could easily be the result. The hateful environment was psychologically oppressive. The state of Mississippi was radically segregated, and was also fiercely fighting the growing Civil Rights movement spearheaded by Martin Luther King. Ron remembers encountering signs that read 'Whites Only' and 'No Coloureds' - separate bathrooms, drinking fountains and eating establishments for blacks and whites was the norm.

The requirement to stay cloistered on base meant that Ronald had a lot of off-duty time to kill - and he did so by occupying his mind with electronics. He also took advanced courses in computers, electronics, engineering and physics. He worked on his mathematical skills. He plunged into the base library and read every book he could find on the great physicists who had developed the field of quantum mechanics - Einstein, Erwin Schrodinger, Wolfgang Pauli, Niels Bohr and many others.

It all paid off. After his military discharge, Ron easily passed his entrance exams to be accepted as a student of physics at Pennsylvania State University. By 1973, he had worked his way to the top and was awarded a doctorate degree in theoretical physics. He became only the 79th black American to earn a PhD in this field - in fact, he remains the only black physics professor in the United States to this day.

## The Pursuit of Time

Part of Ron Mallett's doctoral dissertation concerned the theoretical possibility of using gravity to reverse the flow of time. After he landed a professorship at the University of Connecticut, he continued to grind away tirelessly at his work regarding the possibility that human beings might find a way to manipulate the very fabric of time. Even so, the highly unusual nature of his work meant that he still had to proceed in a way that was veiled in secrecy.

Writing academic papers on the theoretical possibilities of altering time and gravity was one thing, but actively attempting to achieve time travel in the 'real world' would have been professional suicide within the tight-laced scientific community. Time travel was for science fiction writers and 'nutters'.

Ron's work and story was eventually revealed to the general public, however, when a reporter for the respected *New Scientist* magazine convinced Ron to share the details of his work. Despite reservations, Ron described his theories on time travel, and even boldly suggested that building a time machine might be within the realms of possibility.

Rather than producing a blow to the respectability of his distinguished career, the article caught the public imagination. A series of TV interviews and other programmes followed.

It helped that by the year 2000 most of the scientific community now agreed that, in theory, travelling in at least one direction in time - to the past - was not only possible, but completely consistent with the known laws of physics.

Ron had long contended that if the maths says that something does not violate the basic rules of nature, then there is no stopping mankind from developing instruments that will take advantage of what is possible.

## A Time Machine

The fact is, a few other high-profile physicists had already proposed ways in which a time machine could be built, although each of these ideas were completely beyond practicality. For example, Frank Tipler of Tulane University suggested that if one could build a huge cylinder in outer space and drive it into a rotation, one might be able to set huge volumes of electromagnetic energy into a swirling whirlpool motion. This would have the effect of dragging space-time into looping formation, which in turn would create backward time portals or vortexes into the past.

The only problem? Frank Tipler's time machine would have to be sixty miles long and forty miles across and would require the nuclear energy equivalent of about one hundred suns.

But Ron Mallett now believed he had developed a much more simple and practical way to create gravity-bending time portals using the power of laser beams and light.

## Mallett's Time Machine

Ron Mallett has proposed a machine that would use light energy to open doorways into the past. In 2000, he published a peer-reviewed paper which showed that by circulating a beam of laser lights within the confines of a cylindrical-like object, the light-energy of the lasers could bend and warp the fabric of space-time.

The device would employ four circulating laser beams that would stir an interior space like a 'spoon stirring cream into coffee'. The centre of the vortex would be a 'time tunnel' which would resemble the eye of a vortex a few feet across. If an object passed through this vortex eye, it would come out again at some point in the past.

But there is a catch to the Mallett Time Machine - no one would ever be able to travel back further than to a time when the machine was first switched on. Thus, if you began operating the device on 1 January 2014, this would be the limit to where you could travel into the past. If you kept the machine running for an entire year - to 1 January 2015 - you could step into it and theoretically walk back out into any day that transpired within that year.

Obviously, this would mean no thrilling trips back to the Jurassic Period to see the dinosaurs, or a time jaunt into history to personally witness the Battle of Marathon in ancient Greece. It also would mean no trips to the future unless it be a return trip within the bounds of the time period of the machine's operation.

The agonizing aspect of this for Ron Mallett is that he might actually build a workable time machine after all, but he would still be unable to travel back to pre-1955 for an emotional reunion with his beloved father.

Still, a time machine is a time machine, and if Professor Ron Mallett's numbers and concept are correct, it may not be too long before humanity realizes one of its oldest dreams - the ability to travel through time.

Ron Mallett has said that a prototype of his machine is ready to be built right now. He has asserted that by using today's off-the-shelf technology, he could reasonably cobble together his laser-light tunnel for a mere $250,000, or about £150,000.

He doesn't want to accept funding from just anyone, however. For example, he'll turn away any government or military funding offered to make a prototype of his machine. Ron said that the power to time travel would be too dangerous in the wrong hands - so whomever builds this machine must be an agent of utmost responsibility and with only the purest intention in mind.

Just who or what special organization can meet such requirements will be difficult to determine.

## Critics
Ron Mallett has more than a few critics among his fellow physicists, however. Several have pointed out that while Ron's concept for a laser-driven time machine are on the right track, it still would not work for a variety of reasons.

For one, the space-time which Ron used in his calculations employs a singularity, which is a location where the gravitational field becomes infinite in a way that does not depend on the coordinate system.

This singularity would be required even when the power to the laser system is turned off. Also, a singularity is not the kind of space-time that would be expected to arise naturally if the circulating laser were activated in previously empty space.

Another problem is something called the Chronology Protection Conjecture, which is a theorem that was mathematically proven true by the great Stephen Hawking in 1992. It demonstrates that it should be impossible to create closed time-like curves in any finite region that satisfies the weak energy condition. This is a region that contains no exotic matter with negative energy.

So the bottom line, as Ron's critics contend, is that he would need to go some way to overcome the Chronology Protection Conjecture, which would appear to be impossible without violating the laws of physics.

## Captured the Public Imagination

Ron has said he is aware of the problems posed by colleagues, but is determined to press forward anyway, and still maintains he can build a working time machine within ten years.

His work has captured the imagination of many, including high-profile Hollywood film director Spike Lee, who plans to make a movie about Ron Mallett and his quest to conquer time. Spike Lee purchased the rights to Ron's book, *The Time Traveller: One Man's Mission to Make Time Travel a Reality* which has achieved best-seller status. Money made from the movie may even become a source to bring the Hollywood vision of Ron Mallett's dream to a reality.

If he accomplishes his goal, an African American boy who grew up fatherless and poor on the mean streets of New York City would certainly become the most famous inventor ever to have lived - in any time.

# Conclusion

One thing is clear - the concept of time travel has a special place in the psyche of the human race. People seem to have a natural urge to gain the ability to travel in time.

Why?

For some, it's obvious. There are all those, *'If onlys'*.

'If only I could go back and correct that horrible mistake I made!'

'If we could only go back in time and kill Adolf Hitler before he comes to power, imagine all the suffering that will be avoided!'

'If only I would go back and tell the Titanic to steer clear of that iceberg!'

But there are other appeals as well:

'Wouldn't it be great to go back and actually see the dinosaurs?'

'Imagine being able to have a conversation with Socrates!'

'Wouldn't it be great if I could go back in time and find out the numbers for winning the lottery?'

And then there's the future:

'I wonder what life will be like one hundred years from now, five hundred ... a thousand. Will mankind make it?'

'Wouldn't it be great if I could go into the future and bring back the cure for cancer?'

'Will we ever travel to the stars?'

The appeal of the concept to time travel is not difficult to understand: The desire to know what the future will bring is as natural as wanting to travel backwards and undo the mess we have made of things. But more than anything, the concept of time travel evokes that special feeling that all human beings value almost above all others - a sense of wonder.

***

**We hope you enjoyed these first two books in our Time Travel series. If so, please leave a positive review. It will be much appreciated.**

Printed in Great Britain
by Amazon

34682560R00108